DRILLING FOR WATER

Cranfield University stands as one of Europe's most specialised advanced teaching and applied research centres in the areas of engineering technology and management. The university itself is unique in that most of its courses are run for postgraduates, and subsequently represents one of the largest centres for applied research in Western Europe.

Silsoe College, a faculty of the university is a leading international centre for the application of engineering and management to the agricultural food and forestry sectors.

Drilling for Water

a practical manual

SECOND EDITION

Raymond Rowles

Routledge
Taylor & Francis Group

LONDON AND NEW YORK

First published 1995 by Ashgate Publishing

2 Park Square, Milton Park, Abingdon, Oxfordshire OX14 4RN
52 Vanderbilt Avenue, New York, NY 10017

Routledge is an imprint of the Taylor & Francis Group, an informa business

First issued in paperback 2018

British Library Cataloguing in Publication Data

Rowles, Raymond
    Drilling for Water:Practical Manual. –
    2Rev.ed
    I. Title
    628.114

Library of Congress Cataloging-in-Publication Data

Rowles, Raymond
    Drilling for water: a practical manual / Raymond Rowles.
      p.   cm.
    Originally published: Bedford UK: Cranfield Press, 1990.
    Includes index.
    ISBN 1-85628-984-2 (pbk.) : £20.00 ($34.95 US: est.)
    1. Wells. 2. Boring. I. Title.
TD412.R65   1995
628.1'14–dc20

    94-37923
    CIP

ISBN 13: 978-1-85628-984-9 (pbk)

Contents

Acknowledgements

"Geology"
Mr. A. Raistrick.

American Petroleum Institute.

British Drilling Association,
Mr. B.D. Johnson.

Commercial Hydraulics Ltd.

Coredrill Ltd.,
Mr. M.D. Foster.

Dando Drilling Systems Ltd.

Drilco
Smith International Inc.

Drill-Aid Ltd.,
Mr. B.J. Higgins.

Drillquip Ltd.,
Mr. W. Deeming.

Drillquip Ltd. (Hughes Tool Co. Ltd).,

Failing Supply Ltd.,
Mr. James Michael.

Hagglunds-Denison.

Halifax Tool Co. Ltd.
Mr. J.G. Aspdin.

Hands-England Drilling Ltd.,
Mr. B.A. England.

International Association of Drilling Contractors.

SEMAFOR.,
Mr. Pol Lamouric.

Silsoe College,
Mr. Richard Carter.

Introduction

The purpose of this book is to explain as many aspects as possible of drilling a hole in the ground, from the initial choice of equipment, through the cost to the dangers of drilling and some advice on overcoming drilling problems.

The texts in the various sections are confined to water-well drilling and were written specifically with the driller in mind (and all with whom the driller might come into contact); others will follow.

Drilling is a combination of common sense and experience, the former comes first and the latter later. Think everything out, take advice and never get into a situation without first knowing a way out.

Everything that is written here is based on field experience, on actually doing the work described and experiencing all the disappointments and elations alike that come with challenging "mother nature", who doesn't give up her riches lightly.

Two points to remember. Firstly, it is the formation, or formations, to be drilled that determines everything else. From geological information you can decide then, and only then, the type of bit to use, the method of flushing, the drill string to follow the bit and finally the rig upon which all is mounted; no single system is good for all conditions.

Secondly, drilling is not a precise science because you cannot see what you are doing therefore there can be several different ways of doing one job. All this is brought about by experience and you will find endless arguments amongst drilling people on any one point — listen and learn. The best method of doing a job is that which offers the best cost per foot or gallon of water, etc., *overall*.

Just to drill a hole is not good enough, it is the product from the hole that counts whether it be samples of minerals or, as in our case, water — the product should be maximised in quality as well as quantity.

1 Drilling Rig

Foreword to Book One

Let us consider three points of view, each of which will explain the thinking behind what is to follow:-

1) There are drilling machines and machines that drill.
2) We must achieve the lowest cost per foot *overall*.
3) There are no short cuts in drilling.

The amount of research and expertise that goes into the making of a drilling rig (machine) is by necessity vast. The rig must be versatile, or "combination" as it is known, must handle the drilling tools required to do the work in hand and make a profit for the contractor (or win as much water as possible for the aid programme), bearing in mind the necessity of drilling a water well as quickly, efficiently and profitably as possible.

It is not difficult to make a machine that drills a hole in the ground — but at what cost? Many manufacturers of machines that drill who have had little or no idea of the function of drilling (not making) have led the unwary purchaser into buying something which has only resulted in both of them coming to an unfortunate end in these rather difficult economic times.

It is up to the drilling engineer to guide the purchasing people through the maze of avenues that must be negotiated to arrive at a profitable solution.

A case in point would be a drill bit. A second quality bit is cheaper than the first quality but the latter will have a longer life, thus the cost per foot of the bit should be less. This applies to muds, tools, vehicles, the rig and just about all associated equipment — even personnel, who can be regarded as your most important "equipment".

"Short cutting" has sent more contractors to oblivion than anything else. From "stretching" a rig's capacity to where the tools can no longer be pulled to attempting to drill hard rock by the rotary method with a light rig — disaster. Just to get a contract? If that is the attitude not only will the contractor go out of business but the industry will get a bad name.

Perhaps what follows will help.

2

The Drilling Rig

This is probably the most maligned of all drilling tools. You seldom hear anyone say they have such and such a drill string, or such and such a bit, always such and such a rig, but a rig is just something to hang your drilling tools on; it's the tools that do the work but the rig gets the blame for tool failures.

A drilling rig is a crane that turns its weight of tools. Many have become so complicated that they should seldom be considered for work in an environment where workshop facilities are less than ideal. Careful planning and authoritative discussions are the by-words when entering into the purchase of a rig.

In this section we will consider the older type of rotary table rig, the more modern and in most cases more efficient top-drive units and the much older and first "mechanical" method of putting a hole in the ground, the cable percussion rigs. We should bear in mind that we are looking at water-well drilling on a global scale, which includes those countries where water-well drilling is carried out in areas where service facilities are "remote" if available at all.

There are six simple questions that need to be asked in making your choice. They are:-

a) Can the rig lift the weight of tools needed to drill the required hole with a safety margin?
b) Has it sufficient torque to turn the tools at the diameters needed and with sufficient speed?
c) Is the flushing system(s) on board (or alongside) capable of keeping the hole clean?
d) Will it negotiate the terrain over which it is to travel?
e) What about spare parts?
f) Can I service it?

Let us now consider each question in turn.

A) *Lifting capacity*. Probably the simplest of all. The manufacturer will quote the weight the rig will lift, say for argument it is 10 tons. We then take 75% of that figure, 7.5 tons, this being the safety factor. Deduct from it the weight of drill collars to be used, say 2 tons, leaving 5.5 tons. (The questions of weight on bit will be dealt with in a separate section.)

Let us presume that your drill pipe weighs 20 kilos per metre. We then divide 5.5 tons (5500 kilos) by 20 kilos and arrive at 275 metres, then add back the length of drill collars which is, say, 15 metres. Your rig will drill to 290 metres. That is presuming that all the following points are covered and that all aspects of later sections are taken into consideration.

B) *Torque*. Can the rig turn the bit? As bits are mainly sold by the inch we will have to revert to British measurements and take torque of 150lbs/ft per inch of diameter of bit plus one third safety factor. Therefore a ten inch bit requires 10 (inches) × 150 (lbs.ft.) × 1.33 (safety factor). We now know that we need 1995 lbs/ft of torque to drill a ten inch hole.

Now speed of rotation. For reasons that will be explained in a later section, rotary speeds for rock bits (see later) when used on water-well rigs tend to be higher than with, shall we say, a major piece of oil equipment, and we know from experience that a peripheral speed (the speed at which a given point on the circumference of the bit will travel) of 250 feet per minute works well. Therefore the formula is:-

$$\frac{250 \text{ feet/min}}{\text{pi} \times \text{d} \div 12}$$

Therefore, for a ten inch bit (say):-

$$\frac{250}{3.14 \times 10 \div 12}$$

Or 95 revs per minute.

These are general guides applicable to rotary drilling and should be treated as such. Although they can be applied to core drilling.

Different rules apply when using a down-the-hole hammer in respect of both torque and speed. With the hammer, torque requirement is minimal but rotary speed is critical. The peripheral speed is approximately 30 feet per minute and the substitution of 30 instead of 250 in the above formula will give the speed for a 10 inch hammer as about 12 revolutions per minute.

C) *Hole cleaning*. The question of pumps and drilling fluids will be covered in later sections and for description in greater depth we should wait until then.

D) *Travelling*. There is a saying in the drilling business — "If its turning its earning" — in other words the more time you spend drilling (turning) the more hole (or water) will be produced and there is no greater loss of time than when moving an inadequate rig.

The best way to illustrate this is to quote a case history where a 15 ton (weight) rig was mounted on a 6 × 2 (three axles but only one driven) truck and not only was there a problem of the live axle breaking but the rig was confined to drilling holes by the side of main roads because traction was impossible across wet or mildly sandy topography — the greatest care should be taken in choosing the right carrier for your work.

E) *Spare parts*. Most manufacturers will have a standard set of spares for a given period, say for one or two years etc.,which would have been compiled over many years of experience. Take their advice but still check it out. Be cautious.

F) *Service*. Can you service it? If we are all honest with ourselves we would have to look long and hard at the facilities available in "our country". Often things have been found wanting even in some of the "sophisticated" areas of the world so those who live in the less "affluent" areas should investigate how easy (or difficult) a piece of equipment is to "look after". "The simpler the better" is something to consider although this must be related to the work in hand.

A point to think about here is the crew and the training thereof. All due consideration should be given to those people — "a happy crew will earn money for you".

Having created our rig specification we must look at the type of rig i.e.rotary table, top drive or cable percussion. For very slim hole work there is also the spindle type (core drill) but as we are looking at water wells of varying diameters, this is best disregarded.

The rotary table type drilling rig

This type of rig was "invented" in England around the middle of the nineteenth century and the principle "exported" to The United States of America to speed up the drilling of oil wells. The majority of the development of this rig was done in that country although the basic concept has changed little.

Figure 1–4A gives a general view of a typical water well rotary table drilling rig showing the mast (derrick) which is a "single", this meaning that it is only high enough to pull (or trip) a single joint of drill pipe. Some of the larger rigs can be classed as "doubles", meaning that they will "rack" stands of two joints of drill pipe in the mast, but it is most unlikely that anything larger (thribble three joints or fourble four joints) will be used outside the oil drilling industry: more of this later.

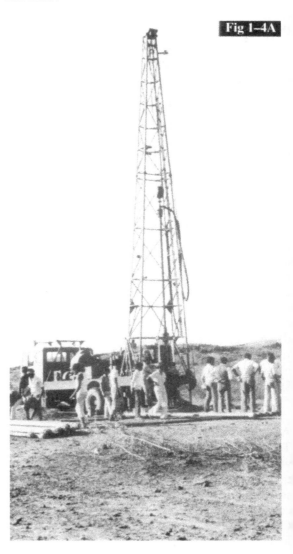

Fig 1–4A

4

At the bottom of the mast is the rotary table which turns the kelly thus transmitting rotary motion to the drill string (column of tools in the hole) and the drill bit.

The kelly and drill string are raised and lowered (fed-off) by the drawworks (a winch or hoist) situated behind the mast.

Whilst some examples of this type of unit are equipped with "crowd" (or pulldown), weight on bit is normally, and quite correctly, applied by drill collars unless, of course, a hammer is used. Under no circumstances should drill collars be used in conjunction with a down-the-hole hammer, otherwise breakage of the hammer will most likely occur.

Fluid is passed from the mud (slush) pump through the kelly hose down the kelly and drill pipe, exiting at the bit to clean the hole carrying cuttings to the surface where the fluid is cleaned and the clean fluid re-circulated — this is known as direct circulation and the fluid can be water or "mud" (more of that later). If compressed air or an air/foam system is used then there is no circulation; once used it is exhausted to the atmosphere.

The mud pump or compressor may or may not be mounted on the rig deck, this being dependent upon the size of hole to be drilled. The larger the hole is in diameter the larger the pump/compressor required.

All components are driven by a prime mover which is usually a diesel engine(s), although electric motors or petrol engines are sometimes used. The engine will usually drive a shaft into a transfer case from which will come the requisite number of shafts (drawworks, rotary table, mud pump etc.) via clutches and brakes which are either actuated manually or by compressed air/hydraulics.

The mast (derrick)

The mast, unlike its top-drive counterpart, takes only the weight of the drill (or casing) string and no lateral torque and does at times seem flimsy compared to the top-drive unit — it should not be underestimated.

The crown block at the top of the mast will carry a number of sheaves over which passes the wireline from the drawworks, thence to the travelling block which in turn carries the swivel, kelly and drill string; the deadman, or free end of the wireline (anchor) is fixed here unless an odd number of lines is used, then the deadman will be attached to the travelling block.

In many cases, where the drill string is "racked" in the mast during a "trip" (pulling the drill string, say, for changing a bit) there will be a racking platform (finger board) which takes the upper part of the drill pipe (collars also) and a monkey board upon which the derrick-man will stand during tripping: taking the string from the hole and returning is known as a "round trip".

The rig chassis

This will carry the rig components and in turn will be carried by a truck, trailer or similar. On very large rigs the chassis will be constructed so as to have working space under it to accommodate blow-out-preventer stacks and high enough to take mud from the hole by gravity into mud tanks, this is known as a sub-structure. Most water-well drilling rigs are mobile and a sub-structure is unlikely.

Figure 1–6A shows a typical chassis layout and in addition to the mast and table it also carries the drawworks, transfer case, engine(s) and often mud pump and compressor.

It is essential whilst drilling to stabilize the rig, i.e. to prevent any movement of the drilling unit, and to effect this the rig chassis is fitted with hydraulic (sometimes manual screw) jacks upon which the rig is lifted, levelled and stabilised, usually on timbers. Experience has shown that lifting the rig high off the ground is unwise, sometimes resulting in chassis bending or total collapse caused by an hydraulic fault. Taking the bounce out of tyres *and* rig levelling are all that is called for.

Some manufacturers offer rigs that can be lifted off the carrier with jacks, thus releasing the vehicle for other uses. Perhaps this could be considered if much time is to be spent over the hole. This is a sort of sub-structure without the capacity of a real sub-structure (figure 1–6B).

Fig 1–6A

Fig 1–6B

Prime mover(s)

The first thing to remember in choosing the rig is to make absolutely certain that all possible doubt is removed concerning servicing and maintenance of your engine(s) in your area. Days or even weeks can be lost for want of a small spare part or a little know-how.

The rig that is totally powered by one engine (all functional parts including mud pump/compressor) is best avoided if the integrity of your operation is to be maintained — perhaps this should be explained and there is no better explanation than with a case history.

The single engined rig in question had an engine failure at the same time as difficulty in the hole. The driller could neither trip his tools or keep his mud circulating. With two engines (one on the rig the other driving the mud pump) he could keep flushing if the rig engine gave out, or make a quick "trip" if the mud pump engine failed — many days were lost firstly in repairing the engine and secondly in "unsticking" his tools. He could have lost the hole and his tools — some have. All this company did was to lose a great deal of money — time costs money. Remember: "when its turning its earning".

6

The transfer case (figure 1–7A)

There is little to say here that manufacturers won't. It is like the transfer box on a large truck, taking power in and transferring it elsewhere. Well maintained, it is quite reliable.

Perhaps a little additional comment should be made here concerning maintenance facilities (or lack of them) in the more remote countries. The rotary table type of rig is easier to "keep going" than the more sophisticated top drive units, although one or two top drive manufacturers have gone to great pains to refute this, with some success. A little bazaar workshop will be able to repair a gearbox but would have doubts about hydraulic pumps.

The drawworks (figure 1–7B)

This is a winch (or hoist) which pulls up or lowers the drill string (or casing), power being taken from the prime mover via the transfer case through a brake which, when drilling, the driller will feed-off manually (although there are rigs equipped with an automatic device), "inching" the drill string down as drilling proceeds. Obviously the whole thing is reversible — it goes up and down.

There can be two drums on the same shaft to accommodate variations in speed and weight requirements; this will be dealt with in a later section.

Fig 1–7A

Fig 1–7B

Fig 1-8A

Cathead (figure 1-8A)

This can be driven off the end of the drawworks shaft or may be driven independently. The cathead is mainly used for lifting heavy pieces of equipment around the rig but may also be used as a spin line for rotating drill-pipe etc in a makeup or breakout situation, whereby a chain is wrapped around the drum of the cathead, thence around the tool to be turned, the helper holding the loose end; thus the tool is turned by operating the cathead.

Auxiliary (bailing) winch (figure 1-8B)

Fig 1-8B

The word bailing is derived from the days of cable percussion drilling when, to clean the hole of cuttings (see later), water was introduced into the hole naturally (groundwater) or manually, and a bailing tool comprising a steel tube with a non return valve on the bottom, was run into the hole on the bailing winch, thus recovering a mixture of water and cuttings. This winch is less powerful than the drawworks but generally faster and is used for auxiliary work around the rig and in the hole other than drilling. It is also known as the sand reel.

Crown block and wire-lines (figure 1-8C)

Fig 1-8C

The crown block is a series of sheaves with an anchor point for the deadman. The choice, care and maintenance of wire-lines are best left to the manufacturer of same, but just be assured that there is sufficient capacity for the work in hand.

The greasing of wire-lines is a point which is much argued over in the industry, but there is one rule that should be remembered. If you are drilling with air flushing, exhaust dust carried from the hole by the air can accumulate on a very greasy wire-line, causing a thickening of the rope by a coating of dust and thus severe damage to the wire-line and the sheaves — be careful.

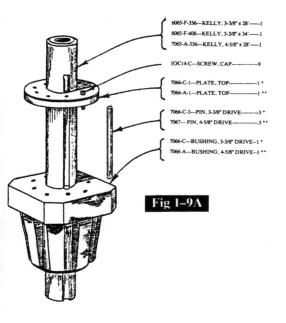

6065-F-336—KELLY, 3-3/8" x 28'——1
6065-F-408—KELLY, 3-3/8" x 34'——1
7065-A-336—KELLY, 4-5/8" x 28'——1

IOC14-C—SCREW, CAP———9

7066-C-1—PLATE, TOP———1 *
7066-A-1—PLATE, TOP———1 **

7066-C-3—PIN, 3-3/8" DRIVE———3 *
7067—PIN, 4-5/8" DRIVE———3 **

7066-C—BUSHING, 3-3/8" DRIVE—1 *
7066-A—BUSHING, 4-5/8" DRIVE—1 **

Fig 1–9A

Travelling block

If drilling was to be carried out off a single line, that line would be connected to the swivel (see later) and depth would be restricted by the capacity of the drum. However, by adding a sheave block (travelling block) with one or more sheaves, and running the wire line around the sheave (or sheaves) and returning to the deadman in the crown block (or travelling block), capacity is increased.

In general terms each additional line will increase the drawworks capacity by that number of lines less about five percent per line for friction, but will reduce the speed at the block by a similar amount. For example a ten ton drawworks (capacity) at 100 feet per minute on a single line will, with two lines "strung" be a little less than 20 tons capacity at the block at 50 feet per minute. The capacity at the block is known as the hook load.

Swivel (figure 1-9B)

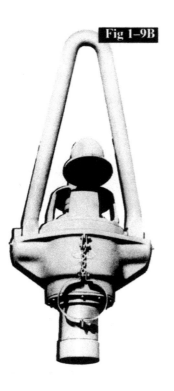

Fig 1–9B

This is a unit through which the mud line from the mud pump is connected to the kelly/drill string via the standpipe. It has three main sections: an outside casing which is stationary and connected to the hook, a rotating inner unit connected to the kelly and — between the two — packings which prevent the leakage of mud — more of that later.

The Kelly (figure 1-9A)

The kelly is connected to the swivel at the top and the drill string at the bottom and is shaped to fit bushings in the rotary table, thus transmitting rotary motion to the drill string. It can be square, hexagonal or fluted and is hollow to allow the passage of drilling fluids. It will be slightly longer than the longest joint of drill pipe, to facilitate the make-up of drill pipe. Uneven lengths of drill-pipe are known as "randoms".

Subs.

Subs. is short for substitutes; otherwise known as adaptors, cross-overs, wear-subs, saver-subs, etc. These adapt between different threads or act as a wearing unit to save the re-threading of expensive items. For instance, at the bottom of, say, a rotary head spindle is a wear-sub of one sort or another because it would be inconceivable to strip down the rotary head and replace the spindle every time the connecting thread to the drill string became worn.

The rotary table (figure 1-10A)

This was given its name to distinguish it from the then only mechanical alternative, the cable percussion rig which had no rotating tools.

The kelly (and drill string) run through the table, and bushings "keyed" into the table and shaped to the kelly give rotational motion to the drill string. Power is either through shafts (or vee belts and pulleys) and pinions from the prime mover or, on more "modern" rigs, from hydraulic motors. Rotation is possible in both directions (forward and reverse).

To make up or break out the drill pipe the kelly and kelly bushings are removed and replaced by taper slips to hold the drilling tools. To run a very large casing the whole table can be removed.

The Top Drive Rig

In water-well drilling the first major impact of top drive rigs was made by the air powered units designed specifically for operating with a down-the-hole hammer. Thus the compressor was the prime mover as well as operating the hammer and cleaning the hole.

In areas where hard rock was near the surface this type of rig made a major contribution to world famine relief. Indeed many old examples of these units are still working and being made. When drilling became more difficult (the softer the drilling the more difficult, the harder the less difficult) the limited torque and power output of the air powered motors made it impossible to cope with these conditions with any degree of confidence; it was then that all knowledgeable heads turned towards hydraulics.

Perhaps we should pause here and ask ourselves a question — the purist geologists say that there is no such thing as overburden — is there? Yes there is.

Overburden is the burden (or weight) sitting on top of your goal. If you are looking for oil at 20,000 feet then you have 20,000 feet of overburden. However, in a drilling context, it has been accepted that "the overburden" refers to the "soft layers at the top", so, to keep matters uniform, the word overburden will be used in the latter context in what follows.

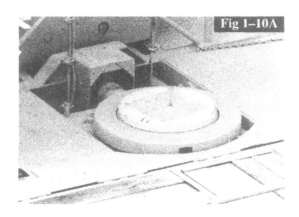

Fig 1-10A

The major part of the development of this type of rig, and its associated hydraulics, came about halfway through the twentieth century when there were louder and louder calls by contractors for a rig that could handle: a variety of rotary drilled diameters; core barrels, without the necessity of "chucking up" after every couple of feet or so; down-the-hole hammers without the annular velocity problems when the kelly gets into the ground and the easing in of longish sections of casing or, indeed, even the "spinning" of same.

Such was the demand for this sort of rig that it seemed to change the concept of "putting a hole in the ground". People hitherto content to sit over a hard rock well for days pumping totally ruinous muds into a water well suddenly realised that there were other ways of doing things. Millions more people can now live a much happier life than they could before this development.

Fig 1-11A

Prime mover(s)

These are almost always diesel engine powered (remember two or more engines are better than one) although some are still petrol or electrically driven. The prime mover will drive one or more hydraulic pumps which pass hydraulic oil via high pressure hoses (sometimes via flow dividers as well) into banks of valves (figure 1-11A) which then divert the oil into one of the rig functions (e.g. rotation, hoist etc.) or back to the hydraulic tank. Once the oil has satisfied the function it too will pass back to tank.

Remember, oil must not come up against a blockage otherwise, at best, it will open relief valves which will overheat the oil, or, at worst, will burst something.

It should be noted that, unlike the rotary table rig, there are no clutches, shafts, gearboxes or brakes to "go wrong", only a ninety percent plus self lubricating power system. However, maintenance of hydraulic components should not be attempted other than in the correct environment.

Perhaps this is a moment to pause and discuss the basics of hydraulics in simple form.

Flow is speed and pressure is force — a basic rule of pumps. An example of this would be, say, a rotary head which is designed to rotate at a variety of speeds and to exert an amount of torque. If it is high speed with low torque, or low speed with high torque, then the required horsepower would be considerably less than a head which produces high speed with high torque (relatively).

Fig 1-11B

11

Hydraulic pumps and motors

Arguably the most controversial and misunderstood parts of the modern flock of top drive rigs by users and makers alike. These should be looked on in the following manner:-

a) They are not user repairable — they must be factory overhauled. Do you have such facilities, or, from the manufacturer's side, are you confident enough in your products or in the country to which they are destined, to guarantee that they will give a long and meaningful life?

b) Are operator skills sufficient to allow the full benefits from the technology presented? Here we can look at a case in point. Some users and manufacturers say that rotation should be infinitely variable from zero to maximum. Firstly we know of no driller or indeed anyone who can honestly give a required rotary speed for any bit in any given formation within a few revs (our speed figures are guides, remember). Secondly, the pumps and motors required to give infinitely variable speed are sophisticated, difficult to overhaul and above all very expensive and not recommended in countries where servicing facilities are unavailable.

c) Will they withstand abuse from unskilled labour for instance when hydraulic oil is topped up? Oil being poured from a dirty tin can has been observed on many occasions — the more sophisticated the unit the more susceptible it is to such abuse.

Types of motors and pumps

Gear Pumps (figure 1-12A). The simplest of all which will take abuse in that they will take a "little" dirt in the oil (not recommended), are easy to maintain, and "cheap". Infinite (?) speed variation can be achieved by using a number of valves and controlling the speed of the prime mover. Another advantage is they can apply shock to a tight bit — others which rely on sophisticated pumps find this near impossible.

The gear type units will tend to "die" over a long period before they become worn out whilst others might "expire" without much warning.

Fig 1–12A

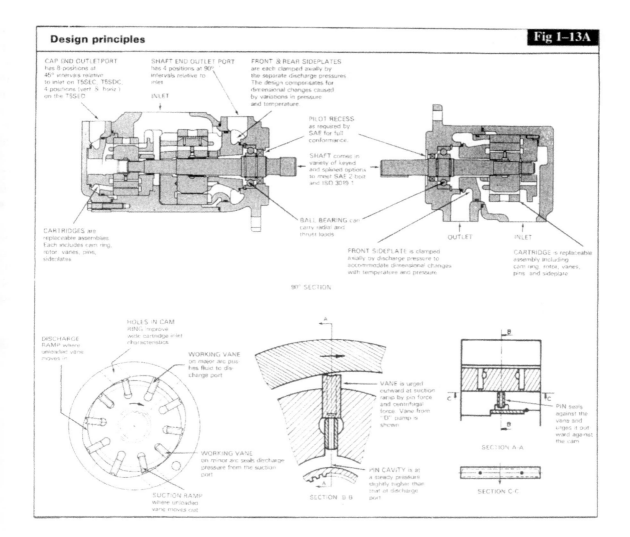

Design principles `Fig 1-13A`

CAP END OUTLET PORT
has 8 positions at
45° intervals relative
to inlet on T5SEC, T5SDC,
4 positions (vert & horiz)
on the T5SED

SHAFT END OUTLET PORT
has 4 positions at 90°
intervals relative to
inlet

INLET

FRONT & REAR SIDEPLATES
are each clamped axially by
the separate discharge pressures
The design compensates for
dimensional changes caused
by variations in pressure
and temperature.

PILOT RECESS
as required by
SAE for full
conformance.

SHAFT comes in
variety of keyed
and splined options
to meet SAE 2 bolt
and ISO 3019 1

BALL BEARING can
carry radial and
thrust loads

OUTLET   INLET

CARTRIDGES are
replaceable assemblies.
Each includes cam ring,
rotor, vanes, pins,
sideplates

FRONT SIDEPLATE is clamped
axially by discharge pressure to
accommodate dimensional changes
with temperature and pressure

CARTRIDGE is replaceable
assembly including
cam ring, rotor, vanes,
pins and sideplate

90° SECTION

DISCHARGE
RAMP where
unloaded vane
moves in

HOLES IN CAM
RING improve
wide cartridge inlet
characteristics

WORKING VANE
on major arc pushes fluid to discharge port

WORKING VANE
on minor arc seals discharge
pressure from the suction
port

SUCTION RAMP
where unloaded
vane moves out

VANE is urged
outward at suction
ramp by pin force
and centrifugal
force. Vane from
"D" pump is
shown

PIN CAVITY is at
a steady pressure
slightly higher than
that at discharge
port

SECTION B-B

PIN seals
against the
vane and
urges it outward against
the cam

SECTION A-A

SECTION C-C

Vane Type Pumps (figure 1-13A). Highly efficient. They will absorb wear in movement of the vanes and will give a long as well as a relatively simple life if looked after. Their fault, if it is a fault, is that they will tend to become unserviceable quite suddenly.

Piston Type (Swash Plate) (figure 1-13B). Probably the most efficient of all which, well looked after in an area where maintenance is available and the crew sympathetic to them, will give a long life. Not ideal though in the less wealthy countries of the world. Questionable from any but the best maintenance. They are the successors to the axial piston pumps which were not variable in flow by the alteration of the angle of a swash plate.

`Fig 1-13B`

The mast (derrick)

This is very much stronger than the mast of the rotary table rig as it not only carries the crown block (cross head), a quantity of hoses, the full torque and weight of the rotary head combined with the weight of the drill string, but almost always the "drawworks" as well.

The construction is so variable that a full book could be written just on that subject. Suffice it to say that it can be box section, single pole, double pole, latticed, pressed steel or any other design that overworked designers think is suitable — "If its strong enough its good enough".

The designed length of the mast is governed by the length of drill pipe that will fit between the underside of the rotary head (*hence top drive*) and the working table at the bottom of the mast. The drill pipe will rarely exceed twenty five feet in length in water-well drilling and the average would be twenty feet (or six metres). The travel of the rotary head must exceed the length of the drill pipe in order to allow for make-up and break-out and to avoid leaving the bit "on bottom" when carrying out these operations.

Attached to the bottom of the mast is the working table (figure 1-14A) which carries a series of bushings acting as guides for centralising the drill pipe, drill collars or hammers during drilling, for the slips bowls and taper slips handling same and, of course, for handling casing.

It is unlikely, although not impossible, that the drill string is racked in the mast, as pipe handling systems have tended to be mechanised since the advent of hydraulics. There are a variety of pipe handling devices, but here we will deal with the three most common:-

Fig 1–14A

A) Carousel (Lazy Susan)

A device attached to the mast into which is loaded a number of drill pipes, similar to the chamber of a revolver, which makes up (attaches) or breaks out (detaches) the pipe to or from the working string. Unfortunately, the carousel cannot handle drill collars.

This is an excellent system, being both fast and accurate, but when the "load" of pipe has been drilled out it becomes cumbersome to reload or, the other way round, unload. There is another similar system where a limited number of drill pipes are actually fitted into the mast — the same problems of loading and unloading occur when the supply is exhausted. Usually a system similar to "c" below takes over when this happens.

B) Side arm loader (figure 1-15A)

Here the drill pipe *and* drill collars are laid down on one side of the rig and are picked up individually (or laid down) by a hydraulic arm. There are no restrictions to the number of tools that can be handled except for the capacity of the rig. This system is fast and assured.

C) Winch and line (figure 1-15B)

A very simple design for a drill pipe handler, but only useful if the rig is equipped with a service winch. It comprises a spring loaded bail attached to one end of a light wire line which has a hook at the other, the wire line being about the length of the drill pipe to be used. About two thirds of a drill pipe length along the wire line away from the hook is a sturdy ring (or clamp) slightly greater in diameter than the drill pipe.

The ring (clamp) is run over the pipe until the hook enters the bottom tool joint of the drill pipe. The winch hook is put into the bail and the pipe lifted by the winch into the mast. When the top tool joint has been made up onto the rotary head the winch is lowered (which in turn lowers the ring), and the handling tool is removed, allowing the bottom tool joint to be made up. Then the handling tool is fitted to the next pipe.

Fig 1-15A

Fig 1-15B

15

Fig 1–16A

The rig chassis (figure 1-16A)

This can be mounted on a truck, trailer, skid, crawler etc., and will carry the mast, prime mover(s), mud pumps, compressor, jacks for levelling or anything else the client (or designer) sees fit, but only within the capacity of the carrier. If a very large mud pump (or, indeed, compressor) is needed then this will usually be a separate unit.

On rare occasions it will carry drawworks, but usually a service winch and, to save weight and space, sometimes the mud pump and/or compressor will be powered by a hydraulic motor.

Please see the safety notes on rig chassis in the section on "The Rotary Table Rig". They apply here also.

Drawworks (otherwise known as hoist)

These are designed to lift the specified weight of tools (drill string) to capacity (remember the safety factor). Feed-off is done hydraulically and can be set to very fine limits indeed, making the use of a variety of tools safe and controlled. The whole system can of course be reversed, not only for reverse movement but also for infinite control of weight on bit by counterbalancing some of the tool weight — an ideal weight on bit is desirable.

There are three main types of drawworks:-

16

A) *Drum*

Similar in most respects to the drawworks of a rotary table rig except that it will usually be powered by hydraulics and will be controlled by same, but in all other respects (down to the travelling block) similar.

B) *Hydraulic motor*

A hydraulic motor of designed capacity is fitted to the top or bottom (or anywhere) of the mast and will drive sprockets and chains which are attached to the rotary head, thus lifting or lowering the head or "feeding-off". A problem here is the surge between "strokes" of the motor motion which can "surge" the bit more and more as depth increases. This should be looked at in the context of depth, and thus as the overloading of the bit (hammer), albeit momentarily.

C) *The hydraulic cylinder*

This is where a hydraulic cylinder, usually through a series of pulleys and wirelines, is connected to the rotary head, giving lift and feed. In this case however, feed on the bit whilst drilling is infinitely controlled by bleeding oil from one side of the cylinder to the other, so there is no surge at any depth within the capacity of the rig. Elemental drilling expertise can "understand" this function very quickly.

The service winch and cathead (figure 1–17A)

With the top drive rig, such a winch should be provided to cope with auxiliary functions such as general tool handling and tidying up the site, but it has two major functions. They are:-

i) *Casing*

Casing a water well is normal and here the installation of a screen or filter is included in the term casing.

In many cases the weight of casing can exceed the drill string weight and your rig choice must take account of this consideration because "running" casing is much more convenient with a winch than off the drawworks. The single drum line pull can be doubled, trebled or quadrupled by adding that number of lines through a travelling block, but remember — firstly, you must allow about five percent loss due to friction on each line and secondly, speed will reduce pro rata to the increase in pull.

ii) *Tripping*

Tripping the drill string is to take it from the hole — round tripping means to take it from the hole and put it back again and so on. In general terms the service winch will trip tools faster than the rotary head, especially at the shallower depths when you have only one line strung over the crown block (also known as a cross head). A hoisting plug (or bail or lifting plug or lifting piece etc.) with the required thread for matching the drill pipe (and collars) is attached to the line and tripping or round tripping commences — time is money.

Fig 1–17A

The cathead (figure 1-18A)

This can be fitted as an extension to the service winch shaft via a clutch or as a separate hydraulically driven unit and is used for general site work or as described in the rotary table rig text.

Breakout Tools (figure 1-18B)

It is general practice with hydraulically powered water-well drilling rigs to mount a hydraulic cylinder on the rig, either in the mast or on the chassis, which will "pull" (and sometimes "push" as well) a tong for breaking out tools. This is a must with modern equipment. This is an area where much time is lost, and a good breakout system will save a great deal of time and, in consequence, money — the contractor *must* aim for the *lowest cost per foot overall*.

Fig 1–18A

Fig 1–18B

The swivel (figure 1-19A)

This is attached to the top or bottom of the rotary head and sometimes, especially for reverse circulation drilling, is through the middle — in other words it is up to the designer of the rotary head where it can be best placed for the work in hand.

The principle is the same as the rotary table rig, where a stationary cylinder encloses one which revolves, separated by packings which prevent loss of fluid.

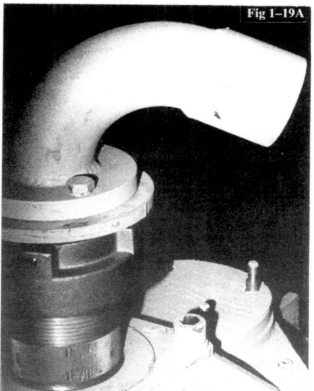

Fig 1-19A

Fig 1-19B

The rotary head (figure 1-19B)

It is impossible to describe a rotary head in detail as there are so many different designs. Needless to say it is entirely at the discretion of the designer to give the best possible arrangement for the work in hand.

The head must be capable of carrying the weight of the designed capacity of drilling tools, and of turning the bits at the correct speeds and torques.

A word of caution here. Rotary heads are very heavy, and that, combined with some inadequate feed-off from the rig or from an inexperienced hand, can damage pipe/collar threads when making-up or breaking-out. Some designers fit a floating spindle in the head, while others offer a hydraulic alternative in order to overcome this problem.

The rotary head should not be held *rigidly* in its carriage otherwise damage will occur to tool joints if there is even the slightest "offset" in spudding (starting) the hole. The threads would then have to be forced together or apart.

Wear-subs etc. have already been described in the rotary table rig and apply here also. However: another word of warning. The top drive rig is often used for working with down-the-hammers and the more this is done the more these words of caution apply:-

As drill pipe threads wear it is not impossible for "feathers" of steel to come off them which are then carried into the hole. In the case of rotary drilling it doesn't matter too much if your mud control is good (you don't want it to go through your mud pump — do you?) but a down-the-hole hammer can be damaged. If you use hammers then drill "pin up" with a box (female) thread on the head facing down — then "feathers" will fall around the outside of the drill-pipe and will go to ground.

The Cable Percussion Rig

Also known as "bash 'n' splash", "walking beam", "cable tool", "jumper" (the latter is incorrect as a jumper rig is something entirely different) and a "spudder". Names given to drilling equipment are sometimes indigenous to a particular area. For instance, a water well is also known as a borehole, a borewell, a tubewell and yet, in the main, it is known as a water well — standard terminology is something we should all strive for.

The history of the cable percussion rig is known to go back about four thousand years and is documented in Chinese history books. The principle is still the same to this day except that it has been mechanised.

A weight (including bit) on the end of a cable is dropped onto the ground to be drilled, thus causing fragmentation and this will be carried on at the discretion of the driller. At a certain point he will pour water (when ground-water is reached that can be used) into the hole and lower the bailer (or sand pump) off the bailing winch, the bailer having a flap valve at the bottom. A mixture of water and chippings will be drawn into the bail, carried to the surface and emptied until the hole is clean — the cycle then continues.

There is, however, another type known as a "shell and auger" rig where the shell (similar to a bailer but designed for cutting) is raised and dropped by a free-fall winch. The auger part of the name is a tube, similar to the shell but with cutting edges. This type of rig is usually quite small and when used to drill water wells, the wells are small and shallow.

There is no mechanical rotation of the tools in either case except for the natural tendency of the wire-line to "turn", which alters the attitude of the bit to the face to be cut, thus achieving a drilled hole. A word of warning here. Make sure the rope mandrill is free to move inside the rope socket otherwise you will set yourself both drilling and mechanical problems. Ensure that your drilling rope has a left hand lay.

The word "borehole" originated to differentiate between a well drilled by this method and one by the rotary system (boring).

In these days of high technology it is hard to imagine why such rigs are still in use. Generally speaking they are slow and cumbersome and mainly purchased for working in areas of low technology (although this is open to doubt) and in soft formations (reasonable thinking) but they are cheap.

Although they might be cheap to buy, in the wrong environment they can be extremely expensive to use in terms of overall cost per foot. Capital cost of equipment represents only a small part of the total drilling cost (see book 9), labour, overheads and fuel being the main contributory factors to the cost.

Confined to soft drilling where one can use such rigs, their ability to run large diameter casings simultaneously with the drilling can prove reasonable, but away from this situation costs increase rapidly. To quote an extreme but noteworthy case history, a cable percussion rig getting into some hard rock was outdrilled more than *ninety* times by a rotary rig running a down-the-hole hammer of similar diameter. Rather than carry around large quantities of drilling casing for percussion drilling it is very expedient to flood the hole with a viscous mud (viscosity keeps the hole open) and drill through the mud, especially in loose formations. You must keep the hole flooded to surface to keep up the hydrostatic pressure and our favoured method is to have a tank of mud on surface and a hose into the hole with a tap at the tank end of the hose. If the level goes down in the hole just open the tap and top up.

Perhaps this is the time to dwell for a moment on another type of rig which seemed a successor to the cable percussion system but never seemed to catch on — although some are even now in use. In the "washdown" or "jetting" rig, a water pump is connected by a hose to the top of a string of steel pipe. Weight is applied to the top of the pipe (sometimes by hands standing on a platform), and rotation is applied by wrenches on the pipe and walking round. At the bottom of the pipe is a "fishtail" bit. The water pump washes out the formation, the weight and rotation do the rest.

And now back to the cable percussion rig and a brief description of its components:-

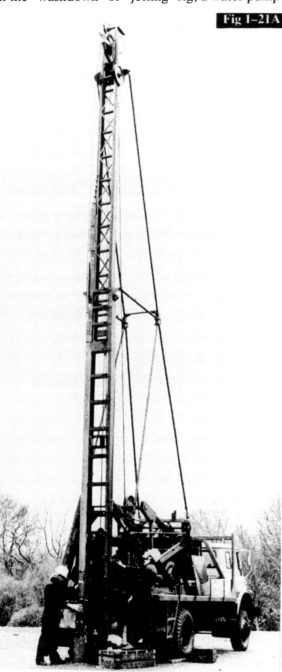

Fig 1–21A

The mast (derrick) (figure 1–21A)

This is usually telescopic, short for transporting and extended for drilling, with long strings of drilling tools and casing. Construction can be latticed, single pole, twin pole etc., and, like the rotary table rig, is designed only to pull or lower — not torque.

The rig chassis (figure 1–21B)

This will carry the winches, cathead and the spudding beam (walking beam), is usually chain or belt driven by a diesel engine (alternatives as usual with all rigs) via clutches and/or brakes, all trailer or truck mounted.

Fig 1–21B

The walking beam (spudding beam) (figure 1-22B)

This is the heart of the rig. A large gear, driven from the engine, drives a pair of cranks and pitman arms which in turn apply the up and down motion to the beam, thereby lifting and dropping the drilling tools.

The bull reel (figure 1-22B)

This usually has a divided drum, one part for storage and the other for working, and is connected to the drilling tools which it lifts and lowers and off which drilling is done.

The calf reel (figure 1-22B)

This is normally used for installation of casing, screens, pumps etc., and will almost always have a travelling block with a couple of lines strung.

The sand reel (figure 1-22B)

This is otherwise known as the bailing winch — used for cleaning the hole with the bailer.

Fig 1–22A

Crown block (figure 1-22A)

Some sort of shock absorber will be there which has two basic functions, firstly to do what its name implies and take shock (stress) away from the structure and secondly to give added "zip" when pulling the bit away from the drilling face at the upward stroke of the beam. Similar in function to or as an addition to "jars" which are described later.

| Controls | Spudding gear | Power unit |

Fig 1–22B

| Countershaft | Spudding beam | Bull reel |

About your Rig

You have your rig and now a few pointers about good housekeeping (rigkeeping?):-

The crew

How many do you need? To be safe and presuming a single shift, you need a driller who can drill. That might sound fundamental but here one must differentiate between a driller and a rig operator. It is not difficult to teach anyone to "pull a few levers", in the course of a day or two with some competence, but you never stop learning how to drill. You can get six different drillers around a table arguing a point and you would get six different answers, all, in their own way, correct and all based on real experience.

You then need two helpers for handling drilling tools and generally assisting the driller.

A fitter/welder is an essential part of any crew and they must be given a comprehensive kit of tools, gas cutting and welding equipment.

Drivers. Always a difficult area but it must be presumed that your driller and fitter each hold a licence to drive, therefore you need one more driver.

Take advice

We all need to do this. If your rig is new then the factory must send an engineer along to train your crew in the operation and maintenance of the rig. If it is second hand then get someone down from the factory to give the rig "the once over" and, again, to train your crew in operation and maintenance.

If you have a drilling problem it is unlikely that the manufacturers themselves would be able to help you so hunt out a drilling engineer; the manufacturer might know of one if you don't.

Support vehicles

To run an operation you need, in addition to the rig, a support truck fitted with a hydraulic arm for lifting heavy loads. Ideally, like your rig, it should be multi axle drive (for maintenance purposes probably a similar truck to that carrying your rig) and this unit will carry your tools around from site to site — drill collars are heavy.

If you are drilling with mud then you will need a tanker (same type of vehicle as above), but if you are operating with an air/foam system then a fair sized tank on the support vehicle is a reasonable alternative. Some support trucks are made with the bed of the truck as a tank.

A crew vehicle is essential, maybe a vehicle similar to a Landrover. This will do the general running about and carry your crew to and from the wellsite. Remember — drilling can be dangerous and a wellsite without a vehicle is a potential disaster area should an accident occur.

Wellsite layout

Lay your site out neatly, it not only looks good but it is safer. Set your drilling tools and the casing on timbers or trestles in the right place for handling into the rig — and neatly.

Have some duckboards around the hole so that the crew are not walking in mud — and make sure the duckboards are cleaned regularly.

Keep hand tools out of the mud and in a place where they can be kept clean and "handy" — you never know when you might need them in a hurry. Make sure any tongs (or wrenches) are well and truly strapped to the rig — how many fishing jobs have been done because a tong or an handtool (or anything) has gone down the hole — and how many holes have been lost?

An essential item for your list is a *First Aid Kit* with a person trained to use same.

2 The Bit

Here it must again be stressed that we are looking at drilling water wells where the rigs are relatively small for the bit sizes they are asked to handle.

Manufacturers' recommended bit loadings for the rockbit go as high as four tons per inch of diameter so, for a six inch rockbit, a loading of twenty-four tons of drill collars should be used.

This means that, allowing for a safety factor (see Book 1) a rig with a thirty ton capacity could only drill with a six inch rockbit to the depth of the drill collars — so what must be done? *compromise* — and that is what this book is all about.

One further point — the correct name for a bit with rolling cones is "a rockbit".

The Bit

The choice of bits is, or has been, so vast that it became imcomprehensible to all but a few. Even that august body "The International Association of Drilling Contractors" (IADC) had to say stop, which they did some years ago, and simplify the choice. But that was only for rockbits which, nowadays, have a more limited usage than in previous times, as will be seen.

What IADC did here was to create a coding for all rockbits across the board, taking in all four of the major manufacturers of first quality bits. Figure 2-25A illustrates this coding system and how to use it — and it must be used as it is now the international language.

Let us consider a major water well at say seventeen and a half inches on our little water well rig. The maker says apply a minimum of two tons to the inch of diameter on your rockbit. That means thirty five tons of drill collars — you should stop shouldn't you, because you only have kelly weight on your rotary table rig and one drill collar on the top drive rig — remember "pulldown" (or crowd) invites disaster. "A bent hole is a spent hole". But the job must be done.

Mostly the overburden (remember — the soft layers at the top?) will be softish so why not use a drag bit? The "weight on bit" for a drag bit can be presumed to be less than half the requirement of a rockbit, therefore you have twice the chance of "making hole" over the rockbit and your pocket is a little fuller because a drag bit is cheaper.

Not only will the humble drag bit drill faster in the softer formations, it will accommodate materials such as clay and gravel better than a rockbit, especially gravels. How many times has the average driller spent hours watching a rockbit turning away on top of a lump of gravel and more than likely doing a lively jig at the same time and making no progress whatsoever? It has been proved over and over again that a drag bit will most likely take the offending lumps, move them between their blades, smash them up and pass the chippings on to higher things.

You have that lovely rig "turning and earning" with a cheap drag bit on bottom and the ground starts getting hard and more than likely you will set casing and cement (see later). Most will remove the drag bit and add a rockbit — commendable and probably all they can do, but the more wealthy or better advised will have a

Fig 2–25A

ormations		IADC Code	1 Non-sealed Bearing	4 Sealed Roller Bearing	5 Sealed Roller Bearing "Heel-Packs"	6 Journal Bearings	7 Journal Bearings "Heel-Packs"
SOFT	1	1	OSC-3AJ	H3A		J1	
		2	OSC-3J	H3		J2	
		3	OSC-1GJ	H1G	HDG	J3	JD3
MEDIUM	2	1	OWV-J	HV	HDV	J4	JD4
		2	WO				
HARD	3	1	W7C			J7	
		2	W7R-2J				
		4				J8	JD8

Example. IADC Code 3-2-1 would be W7R-2J. or Hard/Medium with Non-sealed bearing
Courtesy Hughes Tool Co.

good sized compressor on board and a very nice and fast down-the-hole hammer which will blast just about any rock to Kingdom Come.

The pundits are already shaking their heads and saying what about the annular velocity of the exhaust air? It will be too slow, we need a bigger drill pipe. We stroke our grey beards and shake our aged heads equally gravely and say "what about putting an additive into the air stream to overcome the problem?" — more of that later. What really defeats a hammer is a head of water in the hole — more of that later also.

Is the above a compromise, or is it just common sense? Remember the faster you drill (assuming efficiency) the lower your cost per foot and the greater your return.

We are not saying that the rockbit doesn't have a place in water-well drilling. It does, and always will, because large hammers and compressors are the prerogative of the wealthy (but not smaller hammers and compressors) and there are always those "little bits in between and beyond". Drilling is not a precise science, because you can't see what you are doing — hence compromise and, indeed, the necessity for experience. You can get a dozen different opinions on a specific point all of which might be successful but, which is the most effective in terms of cost per foot overall? That is the crux of the matter.

Let us dwell on the latter point for a moment. If a driller puts holes into, say, a quarry where formations are fairly uniform, he might get problems from time to time, but most holes are foreseeable. The driller who moves around the countryside with a team drilling water wells in a variety of formations, and is successful at it, is vastly more experienced than the one in the quarry. Remember, every time your bit goes below the surface "the thinking starts here".

Having perhaps been a little unkind to the quarry driller in saying that his experience is, perhaps, less than your own, it was this very person who spent years perfecting the down-the-hole hammer because the first commercially viable hammers were designed just for that. Next time you see one, give thanks to them.

What makes a bit work, presuming it is in good condition? There are three things and we are ignoring down-the-hole motors because we have yet to see one working in water wells:-

A) *Rotation*

We have already covered rotary speeds in Book 1, but those of you who would have calculated any other bit size than the one quoted and compared the result with a rockbit manufacturer's leaflets would have seen that our recommended speeds are faster than those. It is our old friend the compromise.

If you look more closely at the rockbit (and drag bit) manufacturers' leaflets, they give a range of weights on bit and a range of speeds and it will be seen that the lower the weight the faster the speed. We can't handle even the minimum weight so we rotate a little faster than the given maximum to compensate, and it works.

Down-the-hole hammers are something else and the guide given in Book 1 to rotation speed for this tool should be looked at.

B) *Weight on bit*

Much has already been said about this but it is always wise to repeat things of importance.

Except in the case of prudent application at the top of the hole, weight on bit should not be applied from the rig (pulldown) but applied with drill collars. A case history:-

A large manufacturer of machines that drill sent their high powered "driller" to the Middle East to drill a fifteen hundred foot hole whilst commissioning a new rig and he did it with pulldown: he completed the hole in record time. The hole was so badly bent, only thirty feet of casing was possible.

Our recommended minimum drill collar weight on rotary bits is one thousand pounds per inch of diameter for rockbits and half that for drag bits. At the top of the hole (where you most need it) this is not possible because there is no drill-string depth to so facilitate this weight on bit, but until such time as you can build towards the desired weight, and as already mentioned above, prudent pulldown is permissible. Be very careful.

The down-the-hole hammer is a different kettle of fish entirely. Weight on bit is quite critical and must be maintained throughout the hole and you must *not* use drill collars.

So what do you do? Well, the weight on bit requirement is quite small and at the top of the hole you can apply the required amount by a little pulldown, but as you get deeper, you will have to counterbalance the excess weight of drilling tools and your rig must be capable of doing this. Mind you, the latter point also applies to rotary drilling with drill collars.

A simple formula for calculating weight on hammer bit is the area of the piston in square inches multiplied by the pressure in psi of air at the hammer. Say the piston area is ten square inches and the air pressure 250 psi then your guide weight on bit is 2500 pounds, but add a little bit to "overbalance". (see Book Nine)

C) *Keeping the bit clean inhole*

This will be dealt with in much greater detail later but perhaps now is the time to mention that cleaning the bit is as important as anything else. "If you can't clean the hole don't start". At best an unclean bit will wear rapidly, at worst it will stick in the hole and you could lose the lot.

"When your engine blows coal, there's trouble in the hole". In other words, a tight bit will send a message back to your rig and the diesel engine will have to take the strain and the exhaust gases will turn black. A good driller drills with ears, eyes and every other sense, including number six.

Another thing to remember is when you are drilling with a liquid (water or mud) there will be a certain amount of flotation of the drill string, which reduces weight on bit. Figure 2-27A is a table of such factors.

Cuttings will settle out in water at the rate of (on average) sixty feet per minute — as mud density increases so this speed reduces. To compensate for this in your flushing fluid minimum annular velocities must be maintained if you are going to drill anything like efficiently. Book Nine contains tables of formulae for calculating these, with examples. The exception to the rule of minimum velocities is *true* air/foam injection drilling where, as long as the column of foam moves, then things are well.

Each of the major bit manufacturers has a team of sales engineers who go around advising customers, and good they are. But each will only offer their company's advice. Try them all — one might have something you really want. Or better still, take advice from a professional drilling engineer. But *always* take advice and remember an expert is a person who knows enough about a subject to realise they don't know it all.

Fig 2–27A

DRILL STRING BUOYANCY

Example of 20,000 pound string in 9.4 lb/gal mud:–
20,000 x .856 = 17,120 lbs

MUD WEIGHT POUNDS/U.S. GAL.	BUOYANCY FACTOR
8.4	.872
8.6	.869
8.8	.866
9.0	.863
9.2	.860
9.4	.856
9.6	.853
9.8	.850
10.0	.847

NOTE. It is dangerous to run drill pipe in compression. To be sure of this it is recommended that you calculate on only 85% of your string weight — this will keep your top 15% of string in tension. Therefore, to let maximum available weight on bit take 17,120 lbs (see above) and divide by 1.15

$$\frac{17120}{1.15} = 14887 \text{ lbs}$$

N.B. Above assumes the use of drill collars.

The Drag Bit

This is the first section as choosing the drag bit should be your first decision.

Drag bits are sometimes called fishtail (because of shape), finger bits, or "Hawthorn" bits. They can be integrated or have replaceable blades, can have two, three or four blades, can be hard steel or tungsten carbide inset, or any combination of these, plus anything else a mind might think up. We have drilled soft sand with a casing coupling which had a blade welded across it and a sub onto that.

Above all else the right choice will be efficient and cheap and you can't beat that in terms of cost per foot.

The drag bit is the granddaddy of all rotary bits, now reborn. It started the whole thing off more than a century ago and in those early days every driller became a designer of bits and every blacksmith's shop resounded to the hammer blows of bits being forged — the drag bit has come back but unfortunately the blacksmith hasn't.

Another more recent design is the replaceable "finger" type bit; the fingers having a tungsten carbide tip and held in place by a "thimble" or circlip. They are a little more expensive than the conventional (is there such a thing?) drag bit but will give rapid penetration into the softer rocks.

The extraordinary speed of the drag bit through the right formations is legendary, but the driller must not get carried away or he can bury the bit and get into all sorts of trouble. Remember "never drill faster than you can clean the hole". Which poses another question. As well as a weight on bit indicator, is your rig equipped with a torque gauge? If not, get one fast, because this is the clearest indication of difficulty in the hole whilst drilling. Let us tell you why with a case history.

The finest method of curing lost circulation problems is to use a foam/polymer/air injection system, because the "stiff" foam will enter the cavities (carrying cuttings with it) until full, then pass to surface. At connection time (adding more pipe) the foam/cuttings will not go back into the hole if used properly. In some areas where the cavities are huge you probably won't get any returns at all so what do you do? Firstly you check to make sure the foam is going into the hole (clear suction hose), make sure air is going with it, and lastly and most importantly, you watch the torque gauge. The slightest build up of cuttings around the bit will result in a flicker of the needle which gets progressively greater as cuttings build up. When this happens you lift the bit off bottom, let it clean up, then carry on drilling, watching the gauge at all times.

In a seventeen and a half inch water well we drilled in the dolomitic limestone of northern Libya, the cavities were so huge we had no returns whatsoever after seventy four feet in a 1000 foot hole, so we decided to drill a ten foot sump at TD (total depth) to take any "fill". We needn't have bothered, our final casing went to our chosen TD and could have gone more.

A point to dwell on here. Never set your final casing and screens on bottom, especially if the screens are of the wire-wound variety. You can collapse them. Natural (or artificial) pack will take care of any support if used properly.

Back to drag bits and when to use them. There are so many weird and wonderful shapes that it seems impossible to recommend one. Just make sure there is a nice big hole (or holes) through which to pass your drilling fluids and that the holes are fairly close to the cutting edge (it is unlikely that your pump/compressor will be enough to jet the hole), then look at the cutter(s).

Cutting action

The action of a drag bit is "paring", or perhaps scraping would be a better word or even shaving. But whatever the word is, the bit is very efficient. Try getting a piece of dryish clay and, with a penknife, cut away small slices, it is very easy to do. With a similar piece of clay try to break it up with a hammer, it is difficult: a down-the-hole hammer hardly drills clay nor indeed will any tool with a percussive action, with any degree of efficiency.

Hard faced steel cutting edge

These are for formations such as sands, clays, shales and the like, bearing in mind they should be of low abrasion.

Tungsten carbide inserts (figure 2-29A)

These will handle the more abrasive formations and can be used in the softer sandstones and limestones and similar rocks. They are available in three or four blades (the latter for slightly harder formations and maintain the gauge better) and will certainly accommodate rocks up to the point where the down-the-hole hammers begin to look attractive.

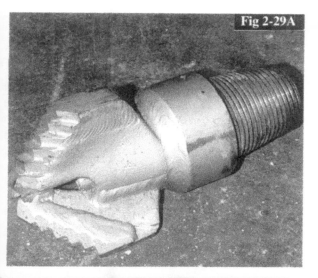

Fig 2-29A

Replaceable blade bits

These comprise connector, locking bowl, mandrill and a set of blades of either type, three or four per set. They may have stabilizers, reamers and the like built in, and are used mainly where you either have a variety of hole sizes to drill or high blade wear. Cutting is, of course, described above for the two integral blade types: they can go up to quite large sizes, and are therefore extremely handy for top-of-the-hole work.

Replaceable finger bits

The finger bit has been with us a long time, especially in augers and is excellent in the more consolidated of the softer formations. Until relatively recently however, they were not exactly readily replaceable, mainly due to the difficulty of holding the "fingers" into the body.

"Trash" in the hole is something to be avoided at all costs. Junking or fishing for "trash" can be time consuming and costly.

However, recent technology has generally overcome this problem and, with the advent of the tungsten carbide tip, the finger bit can be put on your shopping list. Diameters in excess of three feet are not uncommon.

This type of bit falls somewhere between the true drag bit and the rock-bit. Figure 2-29B illustrates an excellent design

Getting to know your bit

Figures 2-30A to 2-30B show the types of drag bits talked about, together with the "names" of the primary components. The names are universal, so you shouldn't have too much trouble communicating any problems to manufacturers.

Fig 2-29B

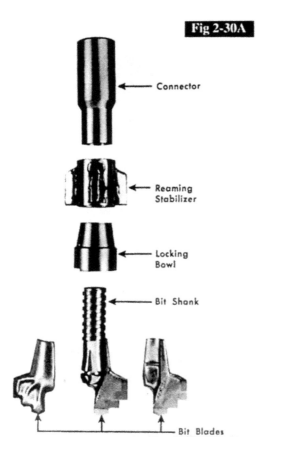

Fig 2-30A

Connector

Reaming Stabilizer

Locking Bowl

Bit Shank

Bit Blades

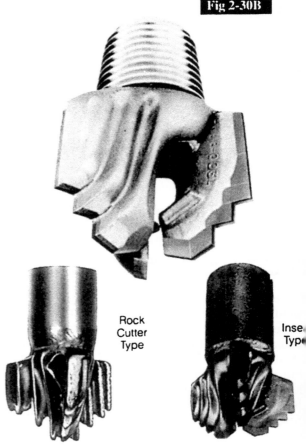

Fig 2-30B

Rock Cutter Type

Inse Typ

Drag bit maintenance

You will need a ring gauge (see figure 2-31A) to check the diminishing diameter of the bit and a set square to check the gauge angle (clearance). It is no good whatsoever having a nice sharp cutting edge that does not cover the full diameter of the hole. For checking gauge angle (clearance) see figure 2-31C.

Make sure your drag bit is sharp along the entire length of the cutting edges. Peripheral dullness will take the load which will not only reduce drilling performance but will cause undue stress laterally and longitudinally (and any angle in between), resulting in the splintering of tungsten carbide inserts (see figure 2-31B).

Keep your bit clean at all times (including when drilling — remember?) and do not let it lie about in the mud.

Use a manufacturers recommended thread dope (grease) at all times.

Never use a hammer to remove the bit from the sub, or to put it back. Get yourself a set of bi breaker plates (see under rockbits for use). A blade can be knocked off with continuou hammering.

Dressing (sharpening) your drag bit i simplicity itself. All you need are the ring gauge set square, and a hand grinder with the correc type of grindstone for the type of bit (steel o tungsten carbide). Then follow the rules liste above.

30

Fig 2-31A

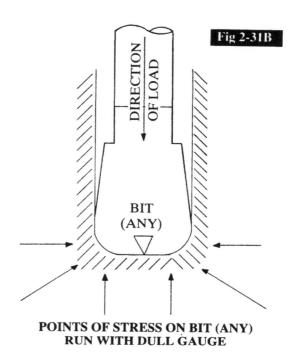

Fig 2-31B

DIRECTION OF LOAD

BIT (ANY)

**POINTS OF STRESS ON BIT (ANY)
RUN WITH DULL GAUGE**

Fig 2-31D

RED BAND

Fig 2-31C

The Rockbit

Also known as the rock roller bit (an unnecessary use of the word roller) or Tricone (a registered trade mark of Hughes Tool Co, which should not be used for general description).

The rockbit has been around for over seventy-five years and in that time has undergone considerable changes. Probably the main change in recent years has been the advent of the tungsten carbide insert bit, which has itself undergone evolution from a design specifically for the drilling of hard abrasive rocks to a complete range from very soft to very hard matching, or indeed, in some cases, having a greater range than the now humble milled tooth bit. But how does that affect us? Not very much really, because weight on bit and r.p.m. are even more critical with the insert bits than the milled tooth, and the cost is prohibitive.

Firstly though, let's look at what a bit does, then perhaps we can get a better understanding. Here we will look at what we will mostly use: the milled tooth bit, and in its three guises, soft, medium and hard formations.

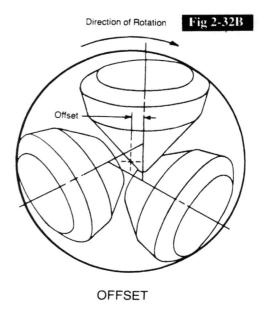

Fig 2-32B

OFFSET

Fig 2-32A

Soft formation (figure 2-32A)

Whilst there are probably two or three grades of soft bit, we will generalise a little, which should be sufficient for our purpose here: this also applies to medium and hard bits, sections on which follow.

There is some crushing action brought about by the drill collar weight (which also affects all other actions); gouging from the rolling motion of the cone which will tend to "flip" a tooth in and out of the ground, thus taking a piece with every tooth and there is scraping which is most desirable in soft ground and similar in function to our old friend the drag bit.

Scraping is there because the centre line of each cone is not perpendicular to the axis of the bit. This is known as offset (figure 2-32B).

The teeth are long and few in number and the bit is used in such things as clay, shale, sands, soft rocks and the like.

Medium Formations (figure 2-33A)

The action is gouging, scraping and crushing/chipping with a greater emphasis on the last mentioned than with the "soft" bit. The offset is less than the soft bit, therefore scraping is less, but what is noteworthy here is the use of the word "chipping".

There are more teeth than the soft bit and they are shorter in length; sometimes a gauge button of tungsten carbide is inserted on the outside of some of the teeth. This reduces the possibility of gauge loss due to abrasion.

Primarily used to drill medium strength formations such as the harder shales, medium sandstones and soft limestones; the starting point of our other old friend the down-the-hole hammer.

Hard Formation (figure 2-33B)

The action is maximum crushing action from your drill collar weight and chipping by the teeth due to the cone rotation.

The gauge teeth are sometimes webbed alternately, thus more resistant to wear. Some will have gauge buttons as well. The teeth are more numerous and shorter than the other types of rockbit. There is no offset.

Let us assume that our drag bit will replace the soft formation bit and the hammer the other two. This isn't strictly true because, even if you have hammers, there are times when a rockbit is very handy. For those without hammers a rockbit is most desirable, nay essential, but which one?

We don't have enough weight on the bit to effect the crushing action and the scraping is limited for the same reason, so what are we left with? Only gouging and chipping.

Now, isn't it reasonable to assume that the more teeth the bit has the better we can maximise these actions, by not only the number of teeth but also by the fact that we rotate at a higher speed (see earlier). Thus more tooth contact per revolution. Again, because we haven't a lot of weight on we can't exactly bury the teeth, so surely we must use a hard formation milled tooth bit where we can't use a drag bit or hammer because of its numerous teeth.

Fig 2-33A

Fig 2-33B

33

Another point to support this. If you look at the bit manufacturers' catalogues you will see that they recommend higher rotation speeds as weight on bit reduces. We don't have the weight but we do have some excess speed.

Here the grey-bearded pundits are shaking their heads in disagreement again. All we can say is that we have yet to see a grey-bearded pundit who has drilled a foot in anger. The above is based on decades of experience and aimed at that very nice person, the driller, who sits out in the heat or cold all day with buckets full of frustrations on one side and ears bashed on the other by grey-bearded pundits.

Back to business.

Let us pause for a moment and ponder the question of bit quality. We have spent years arguing this point with just about every person that has ever been involved in the purchase of drilling equipment. The cheapest is never the best when you are looking at cost per foot overall, but you can change to a system which is cheaper in both cost and in cost per foot, for instance our change from rockbits to drag bits.

The main offenders in this area are, surprisingly, the people who should be looking at saving cost: the people who write or adjudicate tenders, notably the charities or aid programmes. They should remember that, unless something tragic occurs, the longer the bit is "on bottom" and drilling, the cheaper it becomes in cost per foot and it will be found that a first quality bit will give you more "bottom hole" time than the second quality bit by a considerable margin, and the difference in price is not that much.

A case in point. Get two new bits side by side, one first and the other second quality, and ignoring the obvious visible differences ("every picture tells a story"), check the bearings by rocking the cones backwards and forwards — almost certainly the second quality bit bearings will "shake" — in other words it is already worn out — from new!!!

Another factor here is lost time due to poor bit performance. A bit that is worn out and in the hole has to be tripped, another bit fitted, and the new bit run back in the hole. Let us say the bit is at five hundred feet, on twenty foot joints of pipe, and that you have a good crew who will trip at about one minute per joint.

You have fifty times one (minute), that is twenty-five joints out and the same number back, or fifty minutes plus changing the bit, say thirty minutes. You have lost eighty minutes of rig time. So if the second quality bit gives you half the life of a first quality, you have to make two round trips to one, and that is commercial suicide. See book nine — "costings" formula.

The latter comments apply to all types of bit.

Bit Design — flushing

Ah! How we mourn the passing of the regular water course rockbit (centre flushing), buried it seems forever, by the major bit companies. When our rickety old pump clanked away on the surface passing copious quantities of mud, unchecked through the great big hole in the middle of the regular bit, we were happy — now we have jets and a very unhappy pump. Another complication to add to the long list imposed upon the water well drilling industry by the oilfield orientated bit companies.

OK, so they have made great strides with air flush bits, but what about we who drill with mud whose pumps cannot even be considered when calculating a bit hydraulics programme, or when drilling with an air/foam system? What do we do? — we take the jets out, and leave them out or don't buy them in the first place — good old compromise.

We have actually seen bit tender specifications demanding a certain jet size with the bits when the low pressure mud pump, also in the tender, had maximum liner size and was far too small — think about it.

Many of the advantages (?) of jet bits can be handled chemically anyway — the subject of a later book.

Getting to know your rockbit

Figure 2-35A shows the standard terminology for rockbits, a language spoken universally, so if you have a particular problem, then convey it to the manufacturer using these terms.

There is also a basic reporting procedure for bit wear which will be understood by "the business"

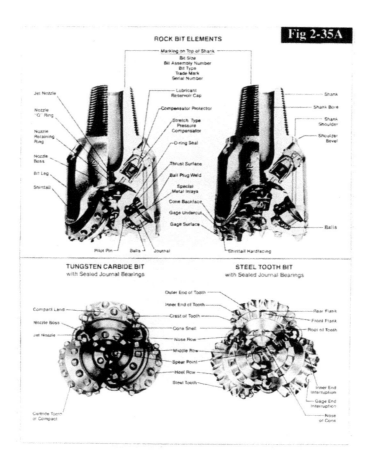

ROCK BIT ELEMENTS — Fig 2-35A

TUNGSTEN CARBIDE BIT with Sealed Journal Bearings

STEEL TOOTH BIT with Sealed Journal Bearings

and figure 2-36A lists that terminology in terms of degrees of wear on the teeth (t), bearings (b) and gauge (g).

In the oil business (bless them) there is a form called a "daily report" which is fairly standard — get the driller to make these out — everyone will gain by it as it has clear and concise details of bit (and other) performance. Figure 2-36A shows the "daily report".

What do you need to get the fullest possible information (and efficiency) from your bit and to get the best possible life from it, other than the absolute care we know you will take when drilling?

Firstly you will need a ring gauge (see figure 2-31A) which will enable you to check gauge wear and, just as important, to check against other bits with which you might have to follow the first. Cone bearings don't like being pinched in a tight hole at all, they tend to react by falling off in

the hole. That is calamitous.

A set square (see figure 2-31C) is a simple yet effective way of seeing how the gauge (clearance) angle (see figure 2-31D) is behaving as the bit gets older; a matchbox can be used if no set square is available.

Never, never hammer the bit out of its sub. Make yourself a set of breaker plates (figure 2-37A) that fit your rig and bit comfortably, and set up your tongs (or wrenches) to suit. *Always* but *always* use a good quality thread dope recommended by the manufacturer on the threads of the bit and every time it comes out of the hole wash it thoroughly and don't let it lay about in the mud.

Do not try to introduce oil into the bearings — it is surprising how such oil will attract sand — and goodbye bearings.

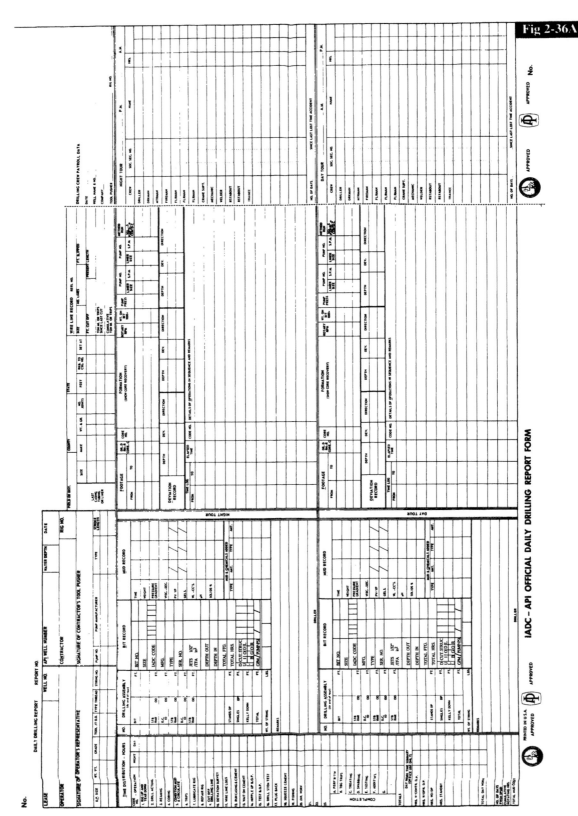

Fig 2-36A

IADC – API OFFICIAL DAILY DRILLING REPORT FORM

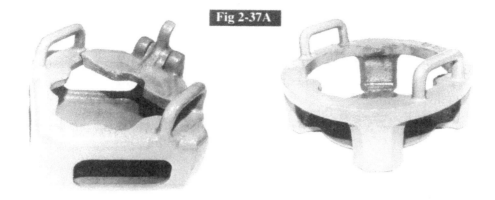

Fig 2-37A

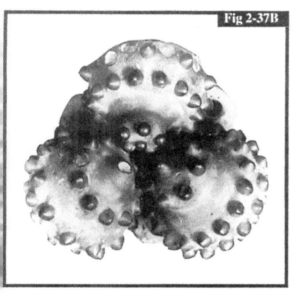

Typical dull condition of a shaped carbide tooth bit.

Tungsten carbide bit with breakage of heel compacts.

The breaker plates should be made to sit in your locked rotary table (or working table on top drive rigs) and the bit should fit snugly into a flame cut shape in the plate. The bit should be unscrewed from its sub with your tong and break out tool and you must also tighten the bit using a similar method. Never put any loose thread in the hole. "An expert fisherman is not necessarily a good driller".

A bit that has been used, whether rockbit, drag bit or hammer bit, will tell a story to the experienced eye and figures 2-37B–C and 2-38A–B are a series of illustrations of different forms of wear with covering notes. You can't pull the wool over experienced eyes.

Before moving on to the the subject of down-the-hole-hammers and bits let us consider other types of bit.

Fig 2-38A

Fig 2-38B

Soft formation steel tooth bit in dull condition showing excessive tooth breakage.

Hard formation steel tooth bit in very dull condition.

Bit Abuse

Please study illustrations 2-37B and C and 2-38A and B. It is rare that a rockbit will fail due to manufacturing faults, it is almost always driller abuse due to a lack of understanding of bit type, weight on bit, rotation speed, flushing medium etc. All of these matters are discussed within this book and if you stick to the basic rules outlined, rockbit failure, or failure of other bit types come to that, will be minimised but don't, repeat, don't be led into buying cheap bits, cost per foot-wise they don't exist.

Don't forget, a bit is only part of the overall drilling operation but a very important part — think about it.

The auger bit

The auger will be dealt with in a later book but we find it interesting to see how some contractors, especially those associated with civil engineering, are beginning to use them more and more for the top of the hole where conditions are known to be soft. If you have a set of augers of the correct size in your armoury, why not use them? After all you don't need a pump or a compressor but you must have enough torque in your rig (see later) and it must be top drive if the augers are of the continuous variety.

Augers are expensive and it is hardly justifiable to spend large sums of money just to save on a bit of "mud". Only use them if you already have them.

Auger bits are usually blade or finger type (already dealt with) or sometimes, with hollow stem augers, a rockbit (wrongly in our opinion) will be used in the centre; you can't be drilling a hole half drag bit and half rockbit — can you?

We suppose that the only two things to mention further are firstly, that the auger bit should be about half an inch bigger in diameter than the flights and secondly, that you should use good quality pins in connecting the bit to the auger and the auger to the auger (or any other efficient method of connection). The rig function of drilling with augers is almost entirely a question of torque (see Book Nine for calculation). You will run out of torque before the drawworks is affected. As hexagon connections wear, the rotation can be thrown onto the pin which, if of poor quality, can be broken or rapidly worn and can easily be pushed out, resulting in a lost string of expensive augers.

If augers are screwed together then they must be supported with some mechanical means of locking the threads, as auger drilling is an operation of forward and, at times, reverse rotation.

Fig 2-39A

39

Under this heading we will also deal with the hole opener's alternative, the large diameter "full face" bit.

Figure 2–39A shows a "full facer" in all its glory: large, beautifully constructed, handsome and weight hungry. We have worked with full facers up to ten feet in diameter, almost always by reverse circulation and at other times with an air/foam system, but always in either civil engineering or mining, never in water wells.

Can you imagine the drill collar weight required by a large diameter "full facer" to push it down, and the torque required to turn it? The purchase of such a beast is a major transaction and they move around from country to country with all the elegance and dignity normally associated with a rig.

So what do we do if we are asked to drill large diameters with our little rigs? We can't even consider a full facer, so we think about drilling a pilot hole and opening it up with... a hole opener.

Hole openers can have as many stages as you like, depending upon your own attitude to the work in hand. They can have rolling cutters or drag blades, and can be direct or reverse circulation. All aspects of torque and weight on bit apply as already listed, but based on the section of hole you are cutting (bearing in mind that you must calculate the area of the ground being cut and not the diameter of the bit because it goes up on the "square"), the area of a twelve and a quarter inch hole is more than twice as much as a six and one eighth hole — it is about four times.

Let us suppose you are asked to drill a twenty-six inch hole and your rig is a bit on the light side for such work. You could drill a twelve and a quarter inch pilot with a conventional bit, pull your tools, put your original bit into a twenty-six inch hole opener to act as a pilot and drill on down with that: instead of your rockbit you can use a bullnose (see figure 2–40A) bit, or plug, (see special note on cleaning the pilot hole later).

Or, if your drill collar weight is restricted, start (round figures) six inches, then twelve, then sixteen, then twenty-six; the alternatives are endless. The calculated choice is yours, but have your mind well and truly set on bit and torque loadings.

You might think this a little too complicated and not worth considering. You know your own country. Can any other company do it more easily? If so, let them do it, but if they can't, such well construction can be very rewarding.

Fig 2-40A

Things to remember about hole openers

Make sure the cutters (roller and drag) are replaceable. The hole opener body should last you for years. Replaceability is sometimes achieved by bolting and tack welding the cutters to a plate — which in turn is bolted and tack welded to the hole opener body. You can have several plates ready drilled and an assortment of cutters to accommodate quite a big range of hole diameters.

Cutters can be simple "bit thirds" (three sections of the rockbit before it is welded up) or complicated, saddle mounted items — there is quite a variety to go at (figure 2–41A).

Take advice from the manufacturer of the cutters as to their placement on the plates at large diameters to get full coverage of the ground to be drilled. It takes some expertise to do this job.

It is unlikely that your mud pump (compressor) will be big enough to clean the hole properly, so it will probably be a case of either reverse circulation and/or an air/foam system. If it is the latter, make sure the foam exits from the pilot bit only — extra holes at the cutters will prevent the foam cleaning the pilot hole (also known as a rathole). When opening a hole, some of the cuttings will fall into the pilot hole, and when your pilot bit reaches them it's got to clean them out.

All aspects of cleaning and maintenance of the hole opener can be taken from the section on drag bits and rockbits.

Cable percussion bits

Here is another name for the cable percussion rig to conjure with — churn drill.

The choice of bit is manifold and is an accumulation of a century or two of experience. They can be long or short or anything in between, chisel or cross, have squarish or rounded shoulders and a host of other things. They have two things in common, they have cutting edges and can be enormously heavy.

In the past the biggest expense in cable tool drilling was for coal to build the fires that heated the bit ready to be attacked by gentlemen with sledge hammers, thereby beating it into a general shape for drilling which they then cleaned up before re-heating for "heat treating".

Today the bit is cleaned up with power grinding or, if in a bad state, built up with special welding rods and finished off with the grinding machine. Let us look at the various aspects of "dressing" the bit:-

Series 8 Cutter

Series 15 Cutter

Series 12 Cutter

Cutting angle

Fig 2-42A

This is a question of choice and the real old-timers in cable tool drilling will have their own preferences but, to generalise, the angle should be between ninety degrees and one hundred and ten degrees, the steeper angle (ninety degrees) for softer drilling and the shallower (one hundred and ten degrees) for the harder formations. It is worth noting that these angles were adopted for the original down-the-hole hammers when they used an all steel cruciform bit at ninety and a tungsten carbide inset at one hundred and ten degrees — soft and hard respectively.

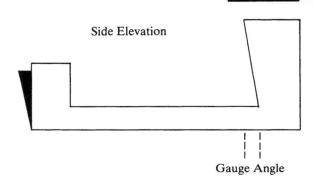

Side Elevation

Gauge Angle

Gauge (clearance) angle

The cutting edge of the bit must cover the entire face (or faces) of the bit terminating in the root of the clearance angle in one continuous edge.

A good rule of thumb is to achieve a gauge (clearance) angle of four degrees from vertical, measured from the extreme outside of the *cutting edge*. If the cutting edge does not cover the complete diameter of the bit, penetration will be impaired and the bit prone to damage.

Like all other kinds of bit you will need a ring gauge to make sure your bit is "in gauge" and a method of checking the clearance (gauge) angle. For the latter a jig similar to figure 2–42A will suffice.

Just for the record, never leave a sharp point on the outside (gauge end) of the cutting edge. Get your grinder and put a flat about one sixteenth of an inch across in its place, otherwise damage can be caused. This applies to the dressing of all types of bits.

As with all bits, they should be treated with care and respect — next time you see a bit have a good look at it. A lot of thought has gone into its development and manufacture — it can make or break an operation — the choice is yours.

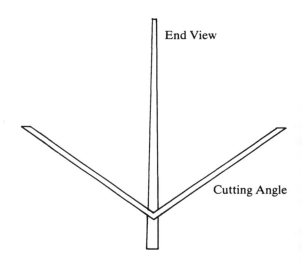

End View

Cutting Angle

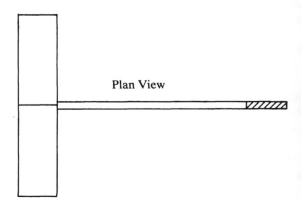

Plan View

Down-the-hole Hammers and Bits

We thought it wise to put hammers and bits together because, whilst we regard the down-the-hole hammer as a high powered set of drill collars, which should be in the next book, others might think otherwise, so we have compromised and put it here.

Controversy has always surrounded this tool and even today it is little used in the oil drilling industry in spite of the fact that there are sizes to suit their drilling programmes. They should no longer have to worry too much about annular velocities of exhaust air cleaning the hole because there are good additives available for this purpose.

A down-the-hole hammer is another tool to hang on your rig for use in the right conditions. That's all it is, not as suggested by some, a panacea for all ills.

A down-the-hole hammer has been around for a little more than thirty years and was primarily designed for the quarrying industry where cable tools were hopelessly slow, as was rotary drilling (although modern techniques have made the latter more acceptable in certain conditions), and the out-of-the hole hammer (drifter) could not drill straight enough to guarantee reasonable calculations of burden and spacing.

There must be a lesson here — drill collars give you a better chance of drilling "straight" holes, so does a down-the-hole hammer. Therefore the latter must be a high powered set of drill collars — QED.

It took some years for even the quarrying industry to trust these tools even though the principle was right because, in the early days there were a tremendous number of holes around the world with broken bits and hammers at the bottom.

But now things have improved and, as far as the water-well drilling industry is concerned, it was the advent of the valveless hammer that really counted, with its ability to absorb fluids together with availability in more appropriate sizes (hammers today range from just below three inches of drilled diameter in diameter to thirty inches).

In the older hammers a valve directed the air above and below the piston and it was this valve that restricted both the ability of the hammer to pass fluids (interruption of movement of valve — flutter) and the use of high pressure air. Let us look at the reasons behind this and how it affected the water-well drilling industry:-

Firstly it must be remembered that pressure of air gives the hammer its speed of penetration. The higher the pressure the higher the performance and the volume of exhaust air cleans the hole. The higher the volume the better the cleaning, but if your hammer consumption of air is restricted by the valve then the drilling speed is limited and, even worse, so is the cleaning of the hole.

An illustration of this. Book Nine is a list of formulae for calculating annular velocities of fluids and from it you will see that the minimum annular velocity of air to clean the hole is 3000 feet per minute and the maximum 5000 feet per minute. In the old days the average water well rig was equipped with three and a half inch drill pipe, which meant it was restricted to a six inch (plus or minus) hammer, because the annular velocity of air was about minimum. But if the driller wanted to drill much bigger holes they had to use bigger drill pipe and for smaller diameters smaller drill pipe. In other words there had to be a second string — not convenient.

Most valved hammers had a by-pass valve fitted. The driller could lift the hammer and the air by-passed the normal valve and more air was put through the hammer to boost the cleaning of the hole. But this was intermittent and the normal air had to be used when calculating annular velocities. Many tried putting "bleed" valves in the string, thus ensuring that air was used up before it got to the hammer. The good drillers could make this work by the use of a number of holes in the "bleed" valve, only using those required, the rest blocked off. Still not terribly convenient.

Attempts were made to pass fluids through the hammer to assist the exhaust air but this upset the movement of the valve, which spoiled the performance of the hammer.

Then came the problem of air pressure. When you are drilling a water well you expect to get water, don't you? Sometimes starting small in quantity and hopefully, increasing to a goodly yield. Under a head of water (column of water in the hole) a pressure is created which increases in direct proportion to the increase in head. So — the air pressure in the hammer is directly affected by this "resisting" pressure and the point comes where the pressure under the head of water becomes the same as in the hammer and the hammer stops. 2.31 feet head of water equals one pound per square inch of pressure.

In can be seen quite easily that a valved down-the-hole hammer with a bottom hole air pressure of one hundred psi would eventually cut out under a two hundred and thirty one feet head of water. Again not convenient and, other than in small, shallow wells (see book one and air powered rigs) the hammer was not a readily acceptable tool for drilling water wells.

Then came the first valveless hammer which worked at higher pressures and consumed greater volume to the point where you had the ability to partially control the throughput volume by a series of chokes. But to the delight of drilling engineers and mud engineers, you could pump fluids through them with consumate ease. Let us pause here and have a look at how this affected the drilling of water wells, with a case history.

It has long been our premise that holes rarely collapse of their own accord. In the majority, collapse is induced by treating the hole wrongly and this never more so than in boulder formations. We believed that in such formations the use of water, mud or air rotary (or hammer) washed away the matrix around the boulders which is normally there, thus the boulder became loose and rolled around causing all sorts of problems.

We were asked to look at such formations with a view to drilling village water wells in a drought area where the boulders were legion and extremely hard. All but one contractor refused to go to the area and the one that did abandoned many holes and those which he did eventually drill took an average of twenty-five days to complete — to only one hundred and sixty feet!

We decided to use an eight inch valveless down-the-hole hammer on the four and a half inch drill pipe the rig had (annular velocity about 2000 ft. per min. — far too low I hear you say) and to inject a mix of drilling foam and polymers. This way the polymers consolidated the matrix while the hammer broke up the hard boulders thus making the hole stable; and the foam overcame the problem of low annular velocity.

Three holes were drilled in five shifts, the last taking one shift (the learning curve) and all cased and screened to depth and successful. A well a day was not impossible.

Back to business. Needless to say, without the fluid capability the hammer is just another tool. With it there is a whole new era being born and all water-well drilling contractors (and a few others) who are worth their salt would be well advised to look at it very carefully.

Now a few words about bits. The cutting action of bits can be related back to the original thinking about cable percussion and rockbits, not forgetting the human race's first thoughts about chiselling away at a piece of anything. If you get the opportunity, look at a good stonemason's kit of tools then look around at any type of drill bit. You can relate the thinking can't you? It is so surprising that it took so long to "get there" — partially due to our species' pig-headedness.

Cable percussion drilling uses a chiselling action (we have already gone into this) and so did the original hammer bits, in that they were chisel type (either cruciform or "X") and even today these are used in softer (non-crushing) formations. But we aren't going to use hammers in such formations are we? No, we are going to use them in their right place, because rotary drilling is far more efficient in softer areas.

Then some bright spark had the idea that crushable formations are best crushed and not chiselled (see remarks on rockbits), and along came the button bit which did just that. Then along came a spate of different length buttons the longer for softer (relative) formations, the shorter for hard (see development of tungsten carbide inset rockbits). Anyway, we now have a goodly selection of button bits available to us.

We won't go into bit body design as that is a book on its own except in two areas:-

The first of these is the method of holding the bit into the hammer: it should be positive. The favourite is the pair of half rings which are held round the top of the bit by an "o" ring above the hammer chuck. The chuck then screws into the hammer and the bit is thus held in place — positive. Perhaps other methods aren't so good so be cautious — fishing a bit out of a hole is expensive. The cost of lost drilling time comes off your profits because costs are ongoing anyway.

The second point is cleaning the hole. There is a thing called "chip hold down", where cuttings are held on bottom by the flushing medium and have to be re-drilled before escaping to top, and this is an ongoing process. It is time consuming and wears your bits out rapidly. Some hammer bits have flushing holes coming through the centre of a plate-type bit with the only escape route being via the buttons — this induces "chip hold down" and bit wear and slow penetration! The more enlightened manufacturers cut great scallops away from the flushing holes or direct the holes out of the sides and centre of the bit close to the cutting face — far and away more efficient — look carefully.

We have already looked at rotary speeds for the hammer (remember — a peripheral speed of thirty feet per minute?) but a couple of basic rules here. The slower you rotate, the faster you drill. The faster you rotate the better you will clean the hole. Both approaches have their snags, especially fast rotation, which will "scrub" your bits out. Our "mean" of thirty feet per minute, peripheral speed, is a good starting place, but if your torque gauge indicates a build up of cuttings around the bit, give the bit a short burst of speed — it works wonders.

An "ideal" rotary speed will deliver the largest average chipping from given formations. Try altering rotation speed (including rotary drilling) and see how this affects overall performance.

Chokes

Most valveless hammers (in our business we should try to forget valves) these days have a centre passage through the hammer by which free or non operational air will pass to assist the exhaust air (operational air) from the piston in cleaning the hole. Into this passage can be placed a series of rubberised bungs. The bungs will either be blank or will have a hole drilled through them, the holes varying in size.

Presuming your compressor has been well chosen and has sufficient capacity for your hammer, you can quite easily calculate from book nine (handy formulae) the amount of flushing air required for your particular hole size against drill pipe size, relate this to the table of choke sizes/air consumption given in the manufactures handbook and bingo you've got yourself a choke size. Slip the choke into the hammer then sit back and enjoy your drilling.

But if you don't have sufficient air available for the work in hand, or you want to conserve fuel cost in you compressor (compressed air is just about the most expensive form of power), then slip in a blank choke which gives basic air volumes to operate the hammer, and inject foam. The foam takes over the cleaning and the air lifts the foam and the hammer will be even happier because most hammers at least, are at their most efficient when drilling with a blank choke.

Shock absorbers

Some hammer manufacturers make a shock absorber which fits on top of the hammer and prevents shock from the hammer being passed up the drill string to the rig; a derivation of a rotary drilling "softshock" which does the same thing and, more interestingly, look at the crown block of a cable tool and see the spring (there is nothing new is there?).

Some people say this protects the drill string, well it does to a certain extent but what it really does is to protect the rotary head/swivel/rotary table or whatever from stress — and that is the main point. If your rotary head is operated via chains and sprockets they need protecting — wire line operation will absorb the shock in the wire lines.

Non return valve (check valve)

This is such an important, nay, almost the most important part of the hammer for water-well drilling. If it works well you wouldn't realise it is there. If it doesn't, and unfortunately some don't, you can waste so much time and money.

Firstly, how it works and why it is there. When you shut your hammer off to make a connection (drill-pipe change) under a head of water, this valve is supposed to close, leaving a charge of air in the hammer which will prevent water and cuttings getting back into the hammer and choking it. If that valve leaks or fails to close you've got yourself a blocked hammer and a round trip on your hands.

This case history refers to rotary drilling where the non return valve is equally important.

The scene is a remote village in India. We are rotary drilling kaolin (very soft and fine grained) and the contractor has failed to supply a rotary bit check valve. We encounter water very early in the hole and block off when changing drill-pipe. After four or five round trips to clean the blocked bit we have not made a foot of hole. We had a hammer which had a "check valve" so, as the kaolin was so soft, we decided to rotary drill with the hammer just to make hole.

After four or five round trips and four or five hammer stripping and cleaning operations because the valve didn't work, we still have not made a foot of hole.

We call a halt and drive to the nearest town and search a dump in a disused workshop where we find a seven eighths of an inch ball bearing and an ancient spring and from these we make a makeshift check valve that slips into the top of the rockbit — from then on drilling proceeds normally.

Fig 2-46A

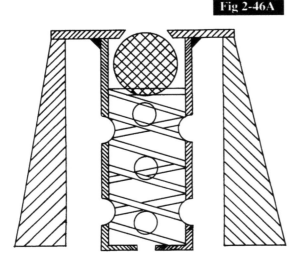

Figure 2-46A is a sketch of that check valve in case you need it.

What is the moral of this story? Read book three where the "in line" check valve is discussed. If your hammer check valve is suspect and you can remove it without causing problems, do so, and fit an "in line" check valve above the hammer. If not, leave it there and still fit the "in line" valve.

Before getting involved in hammer maintenance and the like let us look at something we have already briefly mentioned and upon which we would like to expand — just a little.

The learning curve

If your water-well drilling operation means drilling one hole here, then moving umpteen miles and drilling another and so on and so on then every hole is a challenge, but if you drill a number of holes in an area of similar geology then the first hole will be a challenge, you will learn from that hole to drill the second and from the second to drill the third etc. etc. — that is the learning curve.

Now — what we hope to achieve with these humble offerings of ours is to reduce the length of the learning curve for the latter drillers and to give basic technical assistance to the former so that they can at least start off in a reasonable frame of mind.

Getting to know your Hammer and Bit

Firstly, how the hammer works. Figure 2-47A shows exploded views of a valved hammer and a valveless hammer. They are typical examples but it should be understood that designs differ from hammer to hammer, although the basic principle is the same.

With the valved hammer, compressed air passes through the body of the valve — closing the valve itself — and is directed down the outside of the piston liner between the liner and the hammer casing. The air then goes under the piston which is the hammer of the hammer. This pushes the *piston* on its upward stroke until, at

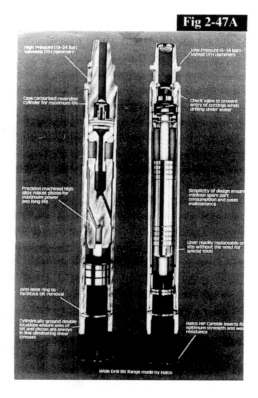

Fig 2-47A

Drilling performance is affected by the pressure and volume of the compressed air used.

Some people say that as pressure increases hammer performance will increase in direct proportion. In fact it increases a little more and is a nice gentle upward curve. For instance, a hammer that drills at twenty feet per hour at one hundred psig will drill a little more than forty feet per hour at two hundred psig.

A hammer drilling at two hundred psig will use about twice as much air in volume as a hammer drilling at one hundred psig, and you need it to clear all those extra cuttings.

Make sure your compressor has ample capacity to handle the hammer's needs with a little bit over the top; that you have good full bore hoses between it and the rig and that the hoses can handle the pressure.

Hammer lubrication

You must do this properly. Intermittent slugs of oil thrown down drill-pipe are no good. Also remember that new drill-pipe will hold oil on its walls, so *overtreat* your hammer in such cases.

Get a good in line lubricator put into the rig that passes the correct pressure and volume of air when picking up oil and that it is adjustable. Make sure you have the correct type and grade of oil to suit your hammer in the ambient temperature of your country. Hammer oil not only lubricates but can close up those little places where air might leak.

The manufacturer of your hammer will give you a list of recommended oils and from the list you will be able to find one which is available in your country. Use it — it will save you a lot of money, and don't forget you will need grease of the correct grade for the threads. The manufacturer should also tell you which.

the top of the stroke, the valve is pushed open and air passes on top of the piston, thus forcing it down and striking the bit. The used air then passes into the hole for cleaning purposes and the cycle starts all over again at an average of 1000 times per minute.

The function of the valveless hammer is very similar except that there is no valve and the compressed air is directed by the piston opening and closing ports in itself, in a centre tube or in the hammer body when it is in motion.

Should we say more? No, not really, it is all as simple as that.

With prudent use of the correct grade of lubricating oil a hammer will wear slowly over a very long period and this wear is quite difficult to detect. The only real indication is a fall off in drilling speed. The main thing that will affect the performance is wear between the piston and liner (or in some cases piston and casing). It is a good idea to get the manufacturer to give you tolerances and to get your hammer checked with internal and external micrometers from time to time.

Hammer maintenance

When not in use wash your hammer through and introduce a film of oil overall. Put a thread protector on the top which closes off the hammer and put some sort of plug in the chuck to stop any dirt getting inside — but don't forget that the plug and thread protector are there when you start again. Also, always test your hammer on surface before running into the hole — its a long way down to bring a hammer back that refuses to "fire up".

Stripping a hammer has always been a difficult proposition and in the past has caused enormous frustrations and cost a great deal in down time. It needn't be so nowadays because simple stripping benches are available that can be used on or around the rig. In some cases they can be powered off the rig hydraulics and others by a simple hand operated hydraulic pump. Please — let us not see too many sledge hammers! Don't forget, the bigger the hammer the bigger the problem.

Fig 2-48A

Bit maintenance

Don't let bits bang together — you will break the carbides. Don't overdrill — the flatter the carbides get the greater chance there will be of inducing fractures. Treat them like very expensive eggs.

We won't dwell on cruciform or "x" bits as they are little used these days but will concentrate on button bits. Grinding cups are available to fit over the buttons (figure 2-48A) — different radius, different cup. These, fitted into the end of a little air powered grinder, will sharpen the carbides up in no time — they need coolant to work efficiently.

You will also need a grinding wheel in case it becomes necessary to cut back the body of the bit just a little.

You will need a set of ring gauges, making it easy to see what wear has occurred and to see if one bit will pass another, and a set of bit breaker plates, all as described in previous chapters.

To reiterate, keep all your tools, and that means hammers and bits as well, out of the mud, sand or anything else that might affect their performance.

Whilst on the subject of bits, make sure you have the correct grade of tungsten carbide for the rock you are drilling. If it is too hard it will break and if too soft will wear out rapidly.

Diamond Bits

There are two primary types of diamond bit, both of which are unlikely to be used in water wells but it is better to look at them and prepare the long suffering driller for what might loom ahead. The first is the full face bit and the other the core bit, but there are also the four following considerations.

1. The division between natural and synthetic diamonds. The latter eminently more acceptable than the former in the right environment.
2. The quality of the diamond which is determined by structure.
3. The size of the diamond, so many per carat, and here we are back to our old friend the rockbit. The larger the diamond (longer the tooth) the softer the formations, the smaller the diamond (shorter the tooth) the harder the formations.
4. Bit design. Here we are back in the hands of the manufacturers, who have to face a drilling problem given to them from the field.

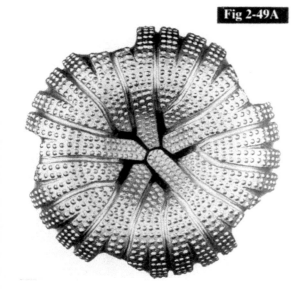

Fig 2-49A

The full face diamond bit (figure 2-49A)

This is a drill bit drilling the complete hole diameter (as opposed to coring) and takes over when there is too great a head of water to allow your hammer to function, when you have gone through the hardest grade of milled tooth rockbit and when tungsten carbide inset rockbits are making slow or no progress. Mind you, if you can afford them (they are vastly expensive), you can use these bits for almost anything. However, most other types of bit are so much cheaper to use in terms of cost per foot overall and in performance it would be foolhardy to even think about it. Their usage is extremely limited.

The action of this bit is to scrape away the rock just as a file scrapes away metal.

When drilling with diamonds, weight on bit and r.p.m. are critical and dependent on the diamond loading and size, and it is very difficult to generalize. So if you should need full face diamond bits, consult a reputable manufacturer. One interesting thing is that the manufacturer will give you T.F.A. (total flow area) for each individual bit to determine mud requirements and this again, is quite critical.

Let us move on.

Diamond core bits

Do you need to core? Yes. Quite often a client, or the local geological survey will ask for one or two cores in a water well to support the chip samples that are normally taken every metre or so. Do you have to take them?

There is no other branch of drilling that creates more mystique than core drilling, and yet it is a normal drilling operation if the rules are obeyed, and need not cause any concern whatsoever unless, of course, a hash is made of it.

It is intensely rewarding when the barrel comes out of the hole and discharges a full core (100%). On the other hand, if there is nothing there it can be a little embarrassing.

We don't intend to dwell too long on core drilling because that is to be the subject of another book, just to give you basic information of how to go about the job:-

1) Make sure the core barrel (figure 2-50B) is in good condition and well "set up". The latter point means that you ensure that the inner barrel (if there is one) rotates freely inside the outer when the barrel is finally assembled and yet the "catcher box" still directs the flushing medium into the bit. A gap of 2 mm between the bottom of the "catcher box" and the bit shoulder seems to be a good measurement to look at. Water/mud flushing is permissible through an "air" barrel but air (or air/foam) can be a problem through a water/mud barrel, so choose carefully.

2) Make sure you have the right type of bit — and it needn't be diamond set. A tungsten carbide inset sawtooth crown does wonders in soft formations and is (relatively) so cheap that it can be regarded as throw-away. Always consult a reputable supplier for diamond bits but if you have any doubt at all about type of bit, and the ground is soft, use tungsten carbide insert sawtooth, and if it is hard and/or badly fractured and mixed, use an "impregnated" bit. Figure 2-50A illustrates various types of core bit.

The "impreg." as it is known, is without a doubt the most versatile and "easy" to use of all diamond core bits. In soft formations it isn't so fast but in mixed formations, where the diamond bit designers scratch their heads, and in hard formations, it will go like the very devil. So there is your choice if you don't have too much coring

Fig 2-50A

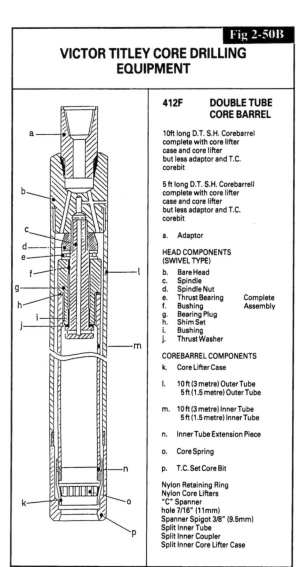

Fig 2-50B

VICTOR TITLEY CORE DRILLING EQUIPMENT

412F DOUBLE TUBE CORE BARREL

10ft long D.T. S.H. Corebarrel complete with core lifter case and core lifter but less adaptor and T.C. corebit

5 ft long D.T. S.H. Corebarrell complete with core lifter case and core lifter but less adaptor and T.C. corebit

a. Adaptor

HEAD COMPONENTS (SWIVEL TYPE)

b. Bare Head
c. Spindle
d. Spindle Nut
e. Thrust Bearing   Complete
f. Bushing   Assembly
g. Bearing Plug
h. Shim Set
i. Bushing
j. Thrust Washer

COREBARREL COMPONENTS

k. Core Lifter Case

l. 10 ft (3 metre) Outer Tube
   5 ft (1.5 metre) Outer Tube

m. 10 ft (3 metre) Inner Tube
   5 ft (1.5 metre) Inner Tube

n. Inner Tube Extension Piece

o. Core Spring

p. T.C. Set Core Bit

Nylon Retaining Ring
Nylon Core Lifters
"C" Spanner
hole 7/16" (11mm)
Spanner Spigot 3/8" (9.5mm)
Split Inner Tube
Split Inner Coupler
Split Inner Core Lifter Case

experience.

By the way, if your "impreg." starts to get polished, a few taps with the edge of a steel rule brings it back into condition.

Use our basic formula for rockbits when it comes to rotary speeds, although a little bit of experiment at slightly higher revs with the "impreg." can be rewarding. Weight on bit can also be experimented with but five hundred pounds an inch is a place to start. Always use a drill collar, especially in mud. The physical design of the core barrel will tend to make it "float" and if you push from the top you will get deviation and, in small diameters, the drill-pipe will bend thus absorbing the imposed load and the drill-pipe will "slap" the hole.

Your bit supplier will have given you the correct water-ways in the bit according to the type of bit and formations, but he should also take into consideration the type of flushing you are going to use.

3) An air bit (compressed air) will normally discharge onto the face of the hole (face discharge) and a liquid bit will discharge behind the "catcher box" and around the "kerf" (cutting edge) which is known as bottom discharge. This is all well and good in stable formations (remember the harder the drilling the easier it is, the softer the more difficult) but if it is unstable then you have got to think again.

4) Figure 2-51A shows the behaviour of fluids around the "kerf" of the bit and it will be seen that your flushing medium is in contact with both

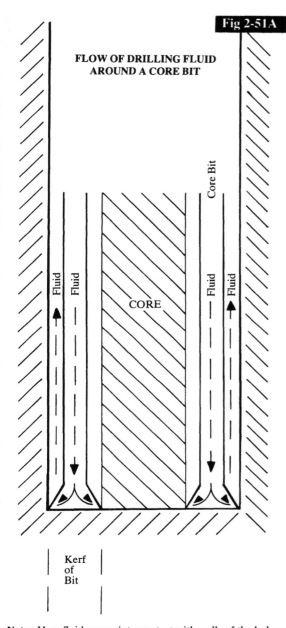

Fig 2-51A

**FLOW OF DRILLING FLUID
AROUND A CORE BIT**

Core Bit

Fluid Fluid

CORE

Fluid Fluid

Kerf
of
Bit

Note: How fluid comes into contact with walls of the hole **and** the core

the wall of the hole and the core as it enters the barrel. If your flushing is either high velocity or high pressure or both you will "washout" any soft sections, and sometimes the soft formation is what you are looking for.

An air/foam/polymer system is almost always the answer here because, as you will have seen before, there is virtually no velocity at the bit, and because the velocity is so low, almost no pressure. This will be dealt with in greater length in a later section but the bit you choose for this application should have large ports and be of the bottom discharge variety; a contradiction for use with air perhaps, but it jolly well works.

It is recommended that the latter bit is over-set by some 3mm on the diameter to allow the foam room to move.

One final "rule of thumb" for coring bits — the coarser the bit the slower the revs, conversely the finer the bit the faster the revs.

3 The Drill String

Foreword to Book Three

In this book you will be introduced to
"standards", things that have not occurred yet,
but things that are pre-eminent in the drill string
because funny things can happen down a hole. In
the old days drillers suffered because no-one ever
calculated the risks. They do now, and in a big
way.

What are standards? They tell manufacturers
that they have to stick to established rules in
things like metal specifications and tolerances,
and they tell contractors which manufacturers to
use.

The British Standards Institute (BSI) have set
out rules on water-well drilling equipment, but
that august body, the American Petroleum
Institute (API) is probably the most famous and
deservedly so — long may they reign.

The Drill String

Before we enter into the body of this offering we would like to emphasise one important point about drill string care that needs to stand on its own, and that is the use of thread protectors (figure 3-53A).

What annoys us immensely is that manufacturers make drill-pipe and other drilling tools and supply them without protectors, or at best offer them as an optional extra. They should be standard on all threaded tools.

You see, a thread protector not only guards your threads, it also stops snakes, spiders and scorpions getting inside tubulars such as drill-pipe and casing and causing mayhem when the tool is used. It also stops muck getting inside your pipe and causing a blockage. Many a time we have heard of a long string of tools being run into a hole only to find the mud wont go down the hole, so they strip the pump or some other such nonsense, then make a trip to find the blocked pipe. What an utter waste of time.

How many times have we asked an inexperienced driller their depth, and seen them fumble in their pocket for a scrappy note book and then start abusing their crew because they don't know either. All tools should have thread protectors and, as we add joints, the thread protectors are neatly stacked on the rig deck and can be counted at any time.

Have we said enough about thread protectors? No, because you can never say enough — they must be used.

We have covered bits already, so in discussing the drill string we will start from above the bit and work our way upwards and that, by the way, is how you should always look at well construction — a moment to ponder that point.

The critical factor in constructing a water well is the size of the pump to be installed (assuming the hydrogeologist knows that the water is there). This goes inside a certain size of casing (and screen), which goes inside a certain size of hole. You have to drill your top hole to accommodate your surface casing which accommodates the hole that accommodates the final casing and the screens which accommodate your pump etc. etc. etc. Obviously, there are times when you have several sizes of casing but you should always start

with your pump size.

There is an API standard for bit and casing clearances (figure 3-54A) — check it out, you might be surprised.

What goes above the bit? — it is a sub.

Fig 3-53A

Substitutes (Subs.)

Also known as adaptors, saver-subs, wear subs, crossover-subs, but collectively, simply subs.

In the drilling business the international terms for describing threads are pin (for male threads) and box (for female threads) and this is how they will be described from now on.

A sub can be pin/pin (male to male) box/box, or box/pin, and that is how it is written. It can also have reamers or stabilizers attached, or a check valve inside, or many other variations on the theme.

Drilling threads are almost always of the API type so, while there are some odd differences with some manufacturers, they are in the minority in the water well industry. We, for our purpose, use only API Standards and the design of threads will be regular (REG.) internal flush (IF), or full hole (FH).

53

Fig 3-54A

Bit Sizes and Clearance – Casing Data

API CASING

CASING SPECIFICATIONS								BIT SIZE		DIAMETRAL CLEARANCE			
Outside Diameter of Casing		Weight with Couplings		Outside Diameter of Couplings		Inside Diameter of Casing				Inches			
Inches	mm	Lbs/Ft	kg/m	Inches	mm	Inches	mm	Inches	mm	Thousandths	Nearest 64th	mm	Nearest mm
4 1/2	114.3	9.50	14.14	5.000	127.00	4.090	103.89	3 7/8	98.42	.215	7/32	5.46	5
4 1/2	114.3	10.50	15.63	5.000	127.00	4.052	102.92	3 7/8	98.42	.177	11/64	4.50	5
4 1/2	114.3	11.60	17.26	5.000	127.00	4.000	101.60	3 7/8	98.42	.125	1/8	3.18	3
4 1/2	114.3	13.50	20.09	5.000	127.00	3.920	99.57	3 3/4	95.25	.170	11/64	4.32	4
5	127.0	11.50	17.11	5.563	141.30	4.560	115.82	4 1/2	114.30	.310	5/16	7.87	8
5	127.0	13.00	19.35	5.563	141.30	4.494	114.15	**4 1/4**	107.95	.244	1/4	6.20	6
5	127.0	15.00	22.32	5.563	141.30	4.408	111.96	**4 1/4**	107.95	.158	5/32	4.01	4
5	127.0	18.00	26.79	5.563	141.30	4.276	108.61	4 1/8	104.77	.151	5/32	3.84	4
5	127.0	21.40	31.85	5.563	141.30	4.126	104.80	3 7/8	98.42	.251	1/4	6.38	6
5	127.0	24.10	35.86	5.563	141.30	4.000	101.60	3 7/8	98.42	.125	1/8	3.18	3
5 1/2	139.7	14.00	20.83	6.050	153.67	5.012	127.30	4 3/4	120.65	.262	17/64	6.65	7
5 1/2	139.7	15.50	23.07	6.050	153.67	4.950	125.73	4 3/4	120.65	.200	13/64	5.08	5
5 1/2	139.7	17.00	25.30	6.050	153.67	4.892	124.26	4 3/4	120.65	.142	9/64	3.61	4
5 1/2	139.7	20.00	29.76	6.050	153.67	4.778	121.36	4 5/8	117.47	.153	5/32	3.89	4
5 1/2	139.7	23.00	34.23	6.050	153.67	4.670	118.62	4 1/2	114.30	.170	11/64	4.32	4
6 5/8	168.3	20.00	29.76	7.390	187.71	6.049	153.64	5 7/8	149.22	.174	11/64	4.42	4
6 5/8	168.3	24.00	35.72	7.390	187.71	5.921	150.39	4 3/4	120.65	1.171	1 11/64	29.74	30
6 5/8	168.3	28.00	41.67	7.390	187.71	5.791	147.09	4 3/4	120.65	1.041	1 3/64	26.44	26
6 5/8	168.3	32.00	47.62	7.390	187.71	5.675	144.14	4 3/4	120.65	.925	59/64	23.49	23
7	177.8	17.00	25.30	7.656	194.46	6.538	166.07	6 1/4	158.75	.288	9/32	7.32	7
7	177.8	20.00	29.76	7.656	194.46	6.456	163.98	6 1/4	158.75	.206	13/64	5.23	5
7	177.8	23.00	34.23	7.656	194.46	6.366	161.70	6 1/4	158.75	.116	7/64	2.95	3
7	177.8	26.00	38.69	7.656	194.46	6.276	159.41	6 1/8	155.57	.151	5/32	3.84	4
7	177.8	29.00	43.16	7.656	194.46	6.184	157.07	6	152.40	.184	3/16	4.67	5
7	177.8	32.00	47.62	7.656	194.46	6.094	154.79	6	152.40	.094	3/32	2.39	2
7	177.8	35.00	52.09	7.656	194.46	6.004	152.50	5 7/8	149.22	.129	1/8	3.28	3
7	177.8	38.00	56.55	7.656	194.46	5.920	150.37	5 7/8	149.22	.045	3/64	1.14	1
7 5/8	193.7	24.00	35.72	8.500	215.90	7.025	178.44	6 3/4	171.45	.275	9/32	6.98	7
7 5/8	193.7	26.40	39.29	8.500	215.90	6.969	177.01	6 3/4	171.45	.219	7/32	5.56	6
7 5/8	193.7	29.70	44.20	8.500	215.90	6.875	174.62	6 3/4	171.45	.125	1/8	3.18	3
7 5/8	193.7	33.70	50.15	8.500	215.90	6.765	171.83	**6 5/8**	168.27	.140	9/64	3.56	4
7 5/8	193.7	39.00	59.04	8.500	215.90	6.625	168.28	6 1/2	165.10	.125	1/8	3.18	3
7 5/8	193.7	42.80	63.69	8.500	215.90	6.501	165.13	6 1/4	158.75	.251	1/4	6.38	6
7 5/8	193.7	47.10	70.09	8.500	215.90	6.375	161.93	6 1/4	158.75	.125	1/8	3.18	3
8 5/8	219.1	24.00	35.72	9.625	244.48	8.097	205.66	7 7/8	200.02	.222	7/32	5.64	6
8 5/8	219.1	28.00	41.62	9.625	244.48	8.017	203.63	7 7/8	200.02	.142	9/64	3.61	4
8 5/8	219.1	32.00	47.62	9.625	244.48	7.921	201.19	7 7/8	200.02	.046	3/64	1.17	1
8 5/8	219.1	36.00	53.57	9.625	244.48	7.825	198.76	6 3/4	171.45	1.075	1 5/64	27.30	27
8 5/8	219.1	40.00	59.53	9.625	244.48	7.725	196.22	6 3/4	171.45	.975	31/32	24.76	25
8 5/8	219.1	44.00	65.48	9.625	244.48	7.625	193.68	6 3/4	171.45	.875	7/8	22.22	22
8 5/8	219.1	49.00	72.92	9.625	244.48	7.511	190.78	6 3/4	171.45	.761	49/64	19.33	19
9 5/8	244.5	32.30	48.07	10.625	269.88	9.001	228.63	8 3/4	222.25	.251	1/4	6.38	6
9 5/8	244.5	36.00	53.57	10.625	269.88	8.921	226.59	8 3/4	222.25	.171	11/64	4.34	4
9 5/8	244.5	40.00	59.53	10.625	269.88	8.835	224.41	**8 5/8**	219.10	.210	13/64	5.33	5
9 5/8	244.5	43.50	64.74	10.625	269.88	8.755	222.38	**8 5/8**	219.10	.130	1/8	3.30	3
9 5/8	244.5	47.00	69.94	10.625	269.88	8.681	220.50	8 1/2	215.90	.181	3/16	4.60	5
9 5/8	244.5	53.50	79.62	10.625	269.88	8.535	216.79	8 3/8	212.72	.160	5/32	4.06	4
10 3/4	273.0	32.75	48.74	11.750	298.45	10.192	258.88	9 7/8	250.82	.317	5/16	8.05	8
10 3/4	273.0	40.50	60.27	11.750	298.45	10.050	255.27	9 7/8	250.82	.175	11/64	4.44	4
10 3/4	273.0	45.50	67.71	11.750	298.45	9.950	252.73	9 7/8	250.82	.075	5/64	1.90	2
10 3/4	273.0	51.00	75.90	11.750	298.45	9.850	250.19	**9 5/8**	244.50	.225	7/32	5.72	6
10 3/4	273.0	55.50	82.59	11.750	298.45	9.760	247.90	**9 5/8**	244.50	.135	9/64	3.43	3
11 3/4	298.4	42.00	62.50	12.750	323.85	11.084	281.53	11	279.40	.084	5/64	2.13	2
11 3/4	298.4	47.00	69.94	12.750	323.85	11.000	279.40	10 5/8	269.88	.375	3/8	9.52	10
11 3/4	298.4	54.00	80.36	12.750	323.85	10.880	276.35	10 5/8	269.88	.255	1/4	6.48	6
11 3/4	298.4	60.00	89.29	12.750	323.85	10.772	273.61	10 5/8	269.88	.147	9/64	3.73	4
13 3/8	339.7	48.00	71.43	14.375	365.12	12.715	322.96	12 1/4	311.15	.465	15/32	11.81	12
13 3/8	339.7	54.50	81.11	14.375	365.12	12.615	320.42	12 1/4	311.15	.365	23/64	9.27	9
13 3/8	339.7	61.00	90.78	14.375	365.12	12.515	317.88	12 1/4	311.15	.265	17/64	6.73	7
13 3/8	339.7	68.00	101.20	14.375	365.12	12.415	315.34	12 1/4	311.15	.165	11/64	4.19	4
13 3/8	339.7	72.00	107.15	14.375	365.12	12.347	313.61	**12**	304.80	.347	11/32	8.81	9
16	406.4	65.00	96.73	17.000	431.00	15.250	387.35	**15**	381.00	.250	1/4	6.35	6
16	406.4	75.00	111.61	17.000	431.00	15.125	384.18	14 3/4	374.65	.375	3/8	9.52	10
16	406.4	84.00	125.01	17.000	431.00	15.010	381.25	14 3/4	374.65	.260	17/64	6.60	7
18 5/8	473.1	87.50	130.22	19.750	501.65	17.755	450.98	17 1/2	444.50	.255	1/4	6.48	6
20	508.0	94.00	139.89	21.000	533.40	19.124	485.75	17 1/2	444.50	1.624	1 5/8	41.25	41
20	508.0	106.50	158.49	21.000	533.40	19.000	482.60	17 1/2	444.50	1.500	1 1/2	38.10	38
20	508.0	133.00	197.93	21.000	533.40	18.730	475.74	17 1/2	444.50	1.230	1 15/64	31.24	31

NOTE. Bit size selections are made from the Hughes "Standard" and "Special Order" size list with a minimum of .014" clearance between maximum bit size (API tolerance) and casing ID — API bit diameter tolerances.

Through 13 3/4" — Nominal to 1/32" (.031")
14" through 17 1/2" — Nominal to 1/16" (.062")
17 5/8" and up — Nominal to 3/32" (.094")

Non-standard bits. All other bit sizes "standard" and normally available.

API drift allowances differing from nominal ID's shown above

8 5/8" and smaller = d − 1/8" (.0125)
9 5/8" to 13 3/8" = d − 5/32"
16" and larger = d − 3/16"

If tolerance appears close, bits should be gaged for clearance

Regular threads (REG)

These are just about always used on rotary bits and have a fairly steep taper (3″ per foot except 6⅝″ which is 2″ per foot) and five TPI (threads per inch) until you get above 4½ inch REG., when these change to four TPI. Regular threads are seldom used on rotary drill-pipe unless flushing is by pure air, because the hole through the thread is quite small and gives resistance (back pressure) in the pipe, thus causing pressure losses which get worse as depth increases. See figure 3-55A for an engineering description.

Internal flush (IF)

This has a shallower taper than REG (2″ per foot), thus the hole through it is larger and will, therefore, reduce friction losses; it has four tpi (see figure 3-55A for engineering description). It is the most widely used thread for drill-pipe, but not always as orginally intended.

Internal flush means a smooth inside bore right through the pipe. Therefore the outside diameter of the tool joint (drill-pipe connections — see later), is greater than the outside diameter of the drill-pipe body (known as external upset — see

All dimensions in inches

Fig 2-55A

			Thread Data				Pin			Box	
Tool Joint Design't'n	Outside Dia. of Pin and Box, ± 1/32 D	Inside Dia. of Pin and Box + 1/64 − 1/32 d	Pitch Dia. of Thread at Gauge Point C	Threads Per Inch	Taper Inches Per Foot on Dia.	Thread form	Large Dia. of pin D_L	Small Dia. of Pin D_S	Length of Pin + 0 − 1/8 L_{PC}	Depth (i) of Box, + 3/8 − 0 L_{BC}	Box Counter-bore. + 0.03 − 0.016 Q_C
REGULAR (REG.) STYLE											
2 3/8 REG	3 1/8	1	2.36537	5	3	V-0.040	2.625	1.875	3	3 5/8	2 11/16
2 7/8 REG	3 3/4	1 1/4	2.74037	5	3	V-0.040	3.000	2.125	3 1/2	4 1/8	3 1/16
3 1/2 REG	4 1/4	1 1/2	3.23987	5	3	V-0.040	3.500	2.562	3 3/4	4 3/8	3 9/16
4 1/2 REG	5 1/2	2 1/4	4.36487	5	3	V-0.040	4.625	3.562	4 1/4	4 7/8	4 11/16
5 1/2 REG	6 3/4	2 3/4	5.23402	4	3	V-0.050	5.520	4.333	4 3/4	5 3/8	5 37/64
6 5/8 REG	7 3/4	3 1/2	5.75780	4	2	V-0.050	5.992	5.159	5	5 5/8	6 1/16
7 5/8 REG	8 7/8	4	6.71453	4	3	V-0.050	7.000	5.688	5 1/4	5 7/8	7 3/32
8 5/8 REG	10	4 3/4	7.66658	4	3	V-0.050	7.952	6.608	5 3/8	6	8 3/64
FULL-HOLE (FH) STYLE											
3 1/2 FH	4 5/8	2 7/16	3.73400	5	3	V-0.040	3.994	3.056	3 3/4	4 3/8	4 3/64
4FH	5 1/4	2 13/16	4.07200	4	2	V-0.065	4.280	3.530	4 1/2	5 1/8	4 11/32
4 1/2 FH	5 3/4	3 5/32	4.53200	5	3	V-0.040	4.792	3.792	4	4 5/8	4 7/8
5 1/2 FH	7	4	5.59100	4	2	V-0.050	5.825	4.992	5	5 3/8	5 29/32
6 5/8 FH	8	5	6.51960	4	2	V-0.050	6.753	5.920	5	5 5/8	6 31/32
INTERNAL-FLUSH (IF) STYLE											
2 3/8 IF	3 3/8	1 3/4	2.66800	4	2	V-0.065	2.876	2.376	3	3 5/8	2 15/16
2 7/8 IF	4 1/8	2 1/8	3.18300	4	2	V-0.065	3.391	2.808	3 1/2	4 1/8	3 29/64
3 1/2 IF	4 3/4	2 11/16	3.80800	4	2	V-0.065	4.016	3.349	4	4 5/8	4 5/64
4IF	5 3/4	3 1/4	4.62600	4	2	V-0.065	4.834	4.084	4 1/2	5 1/8	4 29/32
4 1/2 IF	6 1/8	3 3/4	5.04170	4	2	V-0.065	5.250	4.500	4 1/2	5 1/8	5 5/16
5 1/2 IF	7 3/8	4 13/16	6.18900	4	2	V-0.065	6.397	5.564	5	5 5/8	6 29/64
NUMBER (NC) STYLE											
NC26	3 3/8	1 3/4	2.66800	4	2	V-0.038R	2.876	2.376	3	3 5/8	2 15/16
NC31	4 1/8	2 1/8	3.18300	4	2	V-0.038R	3.391	2.808	3 1/2	4 1/8	3 29/64
NC35	4 3/4	2 11/16	3.53100	4	2	V-0.038R	3.739	3.114	3 3/4	4 3/8	3 13/16
NC38	4 3/4	2 11/16	3.80800	4	2	V-0.038R	4.016	3.349	4	4 5/8	4 5/64
NC40	5 1/4	2 13/16	4.07200	4	2	V-0.038R	4.280	3.530	4 1/2	5 1/8	4 11/32
NC44	6	2 1/4	4.41700	4	2	V-0.038R	4.625	3.875	4 1/2	5 1/8	4 11/16
NC46	6	3 1/4	4.62600	4	2	V-0.038R	4.834	4.084	4 1/2	5 1/8	4 29/32
NC50	6 1/8	3 3/4	5.04170	4	2	V-0.038R	5.250	4.500	4 1/2	5 1/8	5 5/16
NC56	7	3 3/4	5.61600	4	3	V-0.038R	5.876	4.626	5	5 5/8	5 15/16
NC61	8 1/4	3	6.17800	4	3	V-0.038R	6.438	5.063	5 1/2	6 1/8	6 1/2
NC70	9 1/2	3	7.05300	4	3	V-0.038R	7.313	5.813	6	6 5/8	7 3/8

(i) The length of perfect threads in box shall not be less than maximum pin length (L_{PC}), plus 1/8″.

Fig 3-56A

THREAD IDENTIFICATION - API STANDARDS

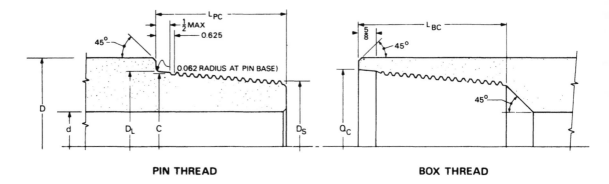

PIN THREAD   BOX THREAD

later), but we in water-well drilling often use a 3½ inch IF thread on a 4½″ drill-pipe, because you get a bigger hole than REG threads and an OD (outside diameter) flush drill-pipe which is a great help when running a hammer, as you avoid interruption of the air coming up the hole. This is similar to internal upset.

Full hole (FH)

As the name implies, this has a full hole for reducing frictional losses. It fell out of favour in the oil drilling industry, probably because it is marginally weaker than IF, but it is an excellent tool for us whose pumps are perhaps a little less efficient than theirs. See figure 3-55A for engineering information.

What has all that got to do with subs you might ask. Well, firstly, subs have threads and secondly, we are about to tell you how to specify one because not many people do (with apologies to those who do).

Let us say you are running a twelve and a quarter inch rockbit which has a six and five eighths REG thread, and for arguments sake, your drill collars have four and a half inch IF in addition to which you are working pin down (remember?) and that is the part most people forget.

Your order would be for a sub screwed six and five eighths API REG. BOX, because all rockbits have pin threads, to four and a half inch API IF BOX. "BOX/BOX".

Let us again suppose that you wanted a new wear-sub between your drill-pipe (say three and a half IF tool joints) and your drill collars. You would order a sub screwed three and a half API IF BOX (from your drill-pipe working pin down) to four and a half API IF PIN (pin down). If, of course, you were working pin up then this sub would be three and a half API IF PIN, to four and a half API IF BOX, wouldn't it? You will see later why we are specifying UP an DOWN.

Even if your drill-pipe and drill collar threads are the same, always use a wear-sub and, indeed do so in any other place where you are mating major tools; it is so much easier and cheaper to replace.

Another thing about subs. They can get a bit tight at times after drilling so ask for key (or spanner) flats to be milled on each side so your tongs can get a bite.

The manufacturer should bore the sub to suit the threads specified and should be at least comparable with the smallest bore of the two threads.

Steel specification is all important, so specify it should be to API Standard. Soft subs will not only wear out quickly but will "gall" and damage your beautiful drill-pipe and collars or whatever they are attached to.

Another pause. The American Petroleum Institute licences certain companies to make certain things and those companies can use the API monogram but only on the items for which they are licensed. These licences are not issued lightly and are jealously guarded by the recipients, so to give yourself some peace of mind, approach these people for your needs; you will be nine tenths of the way there.

What comes next?

Bit Check Valve

Also known as a non-return valve or a "float" valve; it can have an added adjective — "in line". The "float" perhaps needs a little clarification, but we will do that as we go along. It is built into a sub.

The function of a bit check valve has already been described under "down-the-hole hammers" but a little reiteration won't go amiss. When you are drilling under a head of water or mud and you stop drilling to make a connection this valve will close as soon as you shut down your pump/compressor and will prevent any cuttings flowing back into your bit/hammer thus blocking it. It is absolutely essential when drilling with air or air/foam and highly recommended when drilling with mud.

The two main types are mushroom valve (like a car valve including spring) and ball valve. We prefer the former because it can be made much easier than the ball in larger sizes — make sure your valve will pass the required air/mud volumes. See figure 3-57A.

As already mentioned in an earlier section, bit check valves can also be made to fit inside the bit, but you would have to have a number of valves, one for each size of bit, whereas the "in line" is one off — the valves should be replaceable.

These valves will obviously work only in one direction (you put it in upside down and you will block off everything — wont you!) it is, therefore, important to know how to specify the sub in which they work as they will normally lock against a machined shoulder. And this is why we previously specified UP and DOWN when

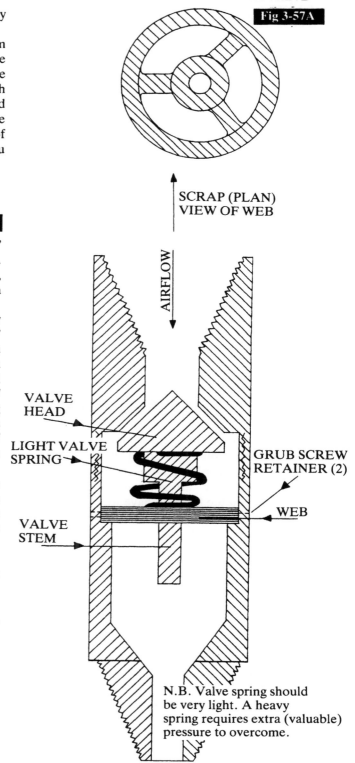

SCHEMATIC OF SELF CLEANING CHECK VALVE

Fig 3-57A

SCRAP (PLAN) VIEW OF WEB

AIRFLOW

VALVE HEAD

LIGHT VALVE SPRING

VALVE STEM

GRUB SCREW RETAINER (2)

WEB

N.B. Valve spring should be very light. A heavy spring requires extra (valuable) pressure to overcome.

57

describing subs — get used to a routine then you wont forget.

The valve is going to fit between your bit sub (it can be incorporated in the bit sub but you have to have one for each size of bit sub) and, let us say, your drill collars, so if you are working pin down with four and a half inch IF threads on your collars you would specify:-

Bit check valve, mushroom type, incorporated in sub screwed four and a half inch API IF BOX UP, four and a half inch API IF PIN down; and give flow rates of fluids and steel specification.

So, why is it called a "float"? You have had a clue already, the fact that it only works in one direction. In the up position the valve is closed, therefore if you run your drill string into a column of water or mud it will tend to float due to resistance from the valve.

This is extremely handy when running heavy strings of casing, not that you would fit this type of valve (unless you were using a cementing shoe — more of that later) you would use some sort of drillable plug (if you have to drill on after setting) with a hole in it, which saves wear and tear on the rig.

Stabilizers and reamers (figure 3-58A)

Although it is possible to incorporate stabilizers or a reamer in the bit sub or "float" we like it here (just above the "float") so, here its going to be; or should we say "they" are going to be — but we haven't told you the function have we?

A reamer is there behind the bit and is usually the same diameter as the diameter of the bit at the start so as to maintain hole diameter as bit gauge wears. It can be straight or spiral, hard faced or tungsten carbide inset or even a roller; or anything else that is in fashion at the time — and you only run one of them and that is in this position. If you dont need a reamer, use a stabilizer here.

And now for stabilizers. They are going to help both your drill collars to keep your hole straight and you to stabilize the length of your drill collars (never your drill-pipe — you'll break them).

They can be roller (soft formations) or blade (harder formations), can be straight or spiral and are usually very slightly less in diameter than your

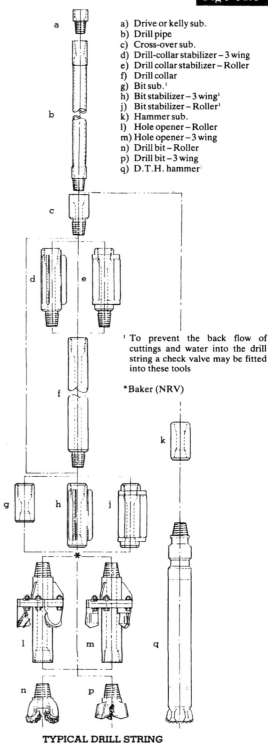

Fig 3-58A

a) Drive or kelly sub.
b) Drill pipe
c) Cross-over sub.
d) Drill-collar stabilizer – 3 wing
e) Drill collar stabilizer – Roller
f) Drill collar
g) Bit sub.[1]
h) Bit stabilizer – 3 wing[1]
j) Bit stabilizer – Roller[1]
k) Hammer sub.
l) Hole opener – Roller
m) Hole opener – 3 wing
n) Drill bit – Roller
p) Drill bit – 3 wing
q) D.T.H. hammer[1]

[1] To prevent the back flow of cuttings and water into the drill string a check valve may be fitted into these tools

*Baker (NRV)

TYPICAL DRILL STRING

bit, maybe a quarter of an inch or so. The overall "effective" (blade or roller) length is about two feet.

If you are running a reamer in the harder, more abrasive formations, because you don't have a hammer, then you wouldn't put a stabilizer behind the bit, but you would put one half way up your drill collars and one at the top of your drill collars and, of course, one behind the bit if you are not running a reamer. Does that make sense? If it doesn't see figure 3-58A

You can now see how important stabilisers are, not only in keeping your hole straight but also keeping your drill collars straight (disaster if not) and away from the side of the hole; they can stick to the side of the hole like glue.

Having, we hope, explained the importance of stabilizers let us give a case history indicating how they should not be used (or specified).

Some manufacturers of "machines that drill" never use drill collars but will sell you a twenty foot long stabilizer (reasonable thinking because it is heavy, but one is not enough). But the way they sell these is not by relating the diameter to the diameter of the bit in use but just, it seems, to any irrelevant diameter.

It has been known for misguided people to sell a nine and seven eighths inch stabilizer behind a twelve and a quarter inch bit, but even worse, for tender documents from a major purchaser of water-well drilling equipment (a charity) actually to specify the same thing — unbelievable!

Let us just have another look at what we have done so far with the drill string with the help of figure 3-58A. We have bit, bit sub, float (check) valve, stabilizer or reamer, a bunch of drill collars, another stabilizer, another bunch of drill collars, then another stabilizer — what next? Another sub onto your drill pipe!

We were tempted to talk about some sort of rotary shock absorber, but this is unlikely to be used in our business at our sort of depths and level of operation.

Just a couple of points here, and they concern the running of a down-the-hole hammer. It has been mentioned several times already but once more wont hurt. You do not run drill collars behind a hammer. The other point, you do not stabilize drill pipe unless drilling anywhere off

vertical, and many water wells are done this way these days.

What we like to see with the hammer is the hammer and bit topped with one stabilizer sub onto your drill-pipe. If your rig has chains and sprockets, then the stabilizer sub would be on top of the shock absorber between that and the drill-pipe.

What about angled or horizontal holes? Well, imagine this:-

If you are drilling such wells (and we have), and you are cleaning the hole very well, your bit and rig act as centres and the deeper you drill the more your drill-pipe will tend to sag, pushing your hole upwards. Now, if you are not cleaning the hole too well, then cuttings will pack under your drill-pipe, pushing your drill-pipe up (drill-pipe is made to flex) and your bit down, and deviation will be downwards.

Under these circumstances the use of stabilizers on your drill-pipe is permissible, but only stabilisers of a certain type. The point is that the bending of unstabilised pipe and the resultant friction with the well can endanger your pipe even more. A calculated risk is better than a ruined hole.

If you have enough space in the hole, then roller type stabilizers (see figure 3-60A) strategically placed up the hole say every twenty feet or so are great, but if you don't have enough space, use non-rotating stabilizers with the same placing — what are they, or what is it?

A non-rotating stabiliser is two bits of tube held together by webs, the OD of the outer tube being about the same diameter as the hole (not bigger) and the ID of the inner tube very slightly bigger than the OD of your drill-pipe. These are run over your pipe and held in place by a shoulder of weld on the outer side (facing out of the hole) and a locking ring on the inner side — or a locking ring on both sides. Your drill-pipe will rotate inside the inner tube and the outer will fit snugly on the hole, causing no damage but keeping it nice and straight. Never use blade type stabilizers — you will tear the hole apart.

Fig 3-60A

APEX – Pipe – Collars, Subs., Stabilizers

APEX DRILL PIPE

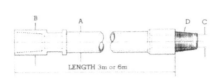

LENGTH 3m or 6m

APEX DRILL COLLAR

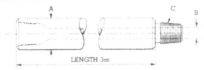

LENGTH 3m

A Pipe o/d in mm O	B Tool Joint o/d in mm	C Tool Joint i/d in mm	D Screwed Connection API		Weight lb/ft kg/m
2⅞ 73.0	3⅜ 85.7	1¾ 44.5	2⅜	IF	7.64 11.37
3½ 88.9	4⅛ 104.8	2⅛ 54.0	2⅞	IF	9.84 14.65
4½ 114.3	5 127.0	2¹¹⁄₁₆ 68.3	3½	IF	13.33 19.83
5 127.0	5½ 139.7	2¹³⁄₁₆ 71.4	4	FH	15.57 23.16
5½ 139.7	6⅛ 155.6	3¾ 95.3	4½	IF	17.00 25.30
6⅝ 168.3	7⅛ 181.0	4 101.6	5½	FH	26.34 39.19

A Collar o/d in mm mm O	B Collar o/d in mm mm	C Screwed Connection API		Weight lb/ft kg/m
4⅛ 104.8	2 50.8	2⅞	IF	34.75 51.72
4¾ 120.7	2 50.8	NC35		49.57 73.77
5 127.0	2¼ 57.2	3½	IF	53.24 79.23
6¼ 158.8	2¹³⁄₁₆ 71.4	4	IF	83.18 123.79
6¾ 171.5	2¼ 57.2	4	IF	108.18 160.99
7 177.8	2¼ 57.2	4½	IF	117.28 174.53
7¾ 196.9	2¹³⁄₁₆ 71.4	NC56		139.28 207.27
8¼ 209.6	2¹³⁄₁₆ 71.4	6⅝	Reg.	160.68 239.12

APEX ROTARY SUBS.

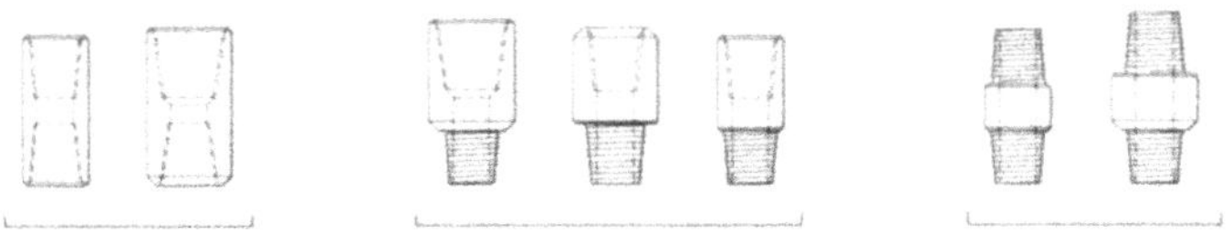

BOX TO BOX

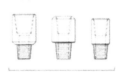

BOX TO PIN

PIN TO PIN

Subs. (an abbreviation for Substitutes) are connecting pieces used to couple together different sizes of drill pipe, drill bits, core barrels etc. or to couple together drill string components incorporating different thread forms.

All internally threaded or female ends are called "Box" and externally threaded or male ends are called "Pin". APEX Subs. are accurately machined from high grade alloy steel.

In order to specify a SUB. correctly the type and size of each thread must be specified. For example a sub. to connect a Drill Pipe having a 2⅜" A.P.I. Internal Flush Pin thread at the lower end to a Drill bit having a 2⅞" A.P.I. Regular Pin thread at the upper end the sub. should be specified as follows:

Sub. 2⅜" A.P.I. — IF Box to 2⅞" A.P.I. — Reg. Box.

APEX DRILL COLLAR STABILIZERS

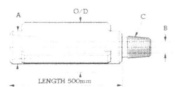

LENGTH 500mm

APEX BIT STABILIZER

LENGTH 500mm

APEX Drill Collar Stabilizers are incorporated in the drill collar section of the drill string. One should be positioned at the top of the upper drill collar with further stabilizers at every 9m (30 ft) down from the top of the collars. An APEX Bit Stabilizer should be installed between the bottom drill collar and the bit.

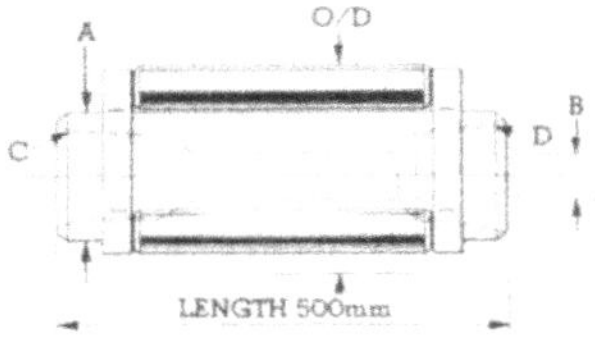

These stabilizers greatly assist in maintaining a straight hole by giving a close fitting and stiff bottom hole assembly.

3 wing stabilizers are for harder drilling whilst roller stabilizers are used for soft formations.

Put the drilling weight on your rotary bits — not pulldown or drill-pipe.

Drill collars are the most frustrating, cantankerous, useless, difficult to handle pieces of equipment out of the hole, but in the hole they make money. Drill collars separate the men from the boys.

Figure 3-61A is a table of drill collars showing, OD, ID, thread form and weight per foot which is really all you need to know about setting up your string to do the work in hand and, of course, your future work. They are a big investment — use the biggest (therefore heaviest) you can over the shortest string that gives you your weight on bit but never leave less than one inch of annulus between them and your bit.

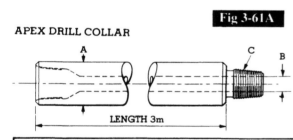

APEX DRILL COLLAR

Fig 3-61A

LENGTH 3m

A Collar o/d in mm mm 0	B Collar o/d in mm mm	C Screwed Connection API		Weight lb/ft kg/m
4⅛ 104.8	2 50.8	2⅞	IF	34.75 51.72
4¾ 120.7	2 50.8	NC35		49.57 73.77
5 127.0	2¼ 57.2	3½	IF	53.24 79.23
6¼ 158.8	2¹³⁄₁₆ 71.4	4	IF	83.18 123.79
6¾ 171.5	2¼ 57.2	4	IF	108.18 160.99
7 177.8	2¼ 57.2	4½	IF	117.28 174.53
7¾ 196.9	2¹³⁄₁₆ 71.4	NC56		139.28 207.27
8¼ 209.6	2¹³⁄₁₆ 71.4	6⅝	Reg.	160.68 239.12

Collars are easier to handle on a rotary table rig than a top drive because the former already has a hole to put them in (you have to bring your kelly back to make a connection, thus you put the collar in the hole and connect the kelly), whereas the top drive demands that you swing the collar into the mast and connect to the rotary head.

All top drive rigs have different ways of handling tools but for drill collars, those where the head tips forward or sideways are better because you can lift one end of the collar up, tip out the head, connect, then run the head up the mast complete with collar. A head that only swings out of the way, say for casing, is not particularly good here because someone has got to go up the mast to connect up — dangerous.

Don't forget the buoyancy factor (figure 2-27A) when calculating your weight on bit, (the heavier the mud the greater the buoyancy), because those heavy tools really do float a little, and don't allow drill-pipe weight to enter your weight on bit calculation — drill-pipe is always in tension, only the drill collar weight is on the bit.

Have you seen the grooves at the root of the drill collar threads yet? No, well have another look at figure 3-61A.

Drill collars are massive, unyielding lumps of high grade steel (big iron) which spin around in the hole, have a tendency to wobble about and, in some formations like fractures and boulders, are thrown up and down in a sort of involuntary hammer action by the reaction of the bit to the formation. This doesn't do the threads much good at all.

The place that movement (or stress) was observed to occur was at the root of each thread (pin and box), and with normally cut threads, such as drill-pipe (drill-pipe bends in its body, not at the thread — if you use good pipe), fracturing took place. The "stress relieving groove" machined into the root a of drill collar largely overcame this problem by spreading the "bending moment" over a greater area.

A good example of this theory is a down-the-hole hammer bit. Next time you see one look at it carefully and you will see that just about every machined angle has a radius and not a sharp angle; in the early days of manufacture bits snapped like carrots at sharp angles.

When you "make up" (connect) your drill

collars make sure you tighten them well, otherwise you will promote failure by accentuating movement, and the same goes for drill-pipe, subs, hammers and anything else that goes down the hole.

We could go on giving advice about drill collars because they are such appealing things — they make money. Never use sharp angled threads (REG for instance) as they will also encourage stress (there are exceptions to this rule in the large sizes), or always use the recommended type of thread dope (grease), or be careful handling them because they are dangerous. But we won't go on because we know you already know all that.

We have already discussed the point that on top of your drill collars goes a sub and stabilizer so lets move on to drill-pipe on our way up the hole.

Drill-pipe

Also known as drill rod (that's diamond drilling), drill stem (that it ain't,), drill steel (that is for drifter or extension steel drilling), drill tube (thats for down-the-hole hammer drilling where you dont need torque) no, gentlemen, its simply drill pipe.

This nonsensical misuse of terms is astonishing. For instance, how many times have you heard or seen the use of the word "barrel" to describe a large steel container of, say, engine oil. The barrel is a measurement which is largely fictitious or at best a theoretical measurement of forty two United States gallons, the drum, which is fifty five United States gallons, is the beast in which you buy engine oil or foam.

The drilling industry has its own language. Just as doctors have to know Latin to do their job, so the driller must know "Drilling". In Book Ten, "A Glossary of Terms" goes some way towards this.

Let us get on with drill-pipe. Wondrous stuff, totally abused, mostly misunderstood and flagrantly misused; and yet, in its long suffering way it will go on earning money until it can no longer take the brick-bats of the uninitiated. How much more it could do if it were well looked after.

How many times have you seen a truck load of drill-pipe arrive on a well-site, without thread protectors, having one end pushed off the truck, then the truck driven away and the whole string crash to the ground in a jumbled heap? We have seen it. Total disaster.

Drill-pipe is specified by the outside diameter of the body of the pipe, followed by the tool joint size, then the length and finally the grade. Let us look at those points in sequence.

1) Outside diameter. Two and three eighths, two and seven eighths, three and a half and four and a half inches are, or at least, have been, the most commonly used in water well drilling, but in recent times the tendency has been towards the last two sizes except where a light rig has to be stretched for reasons of economy.

To illustrate, a bit of mis-use we recently observed.

A major charity supplied a rig to a hard rock area of North-East Africa to drill from nine and seven eighth inches maximum down to six and a quarter inches in diameter. The rig was set up with pulldown but had only one five and a half inch drill collar, no hammer and only two and three eighth inch pipe. You can imagine what happened, the "driller" came upon hard rock, applied pulldown, and ended up with a hole full of spaghetti — that is how it shouldn't be done.

Choose your drill-pipe well. The size (OD) of your drill pipe shall depend upon the size or range of hole diameters you plan to drill. We have already discussed annular velocities of drilling fluids and there is more to come but that is your "yardstick" — always think about your fluids.

Remember, when you calculate your velocities, to take the biggest annulus in the hole. That will usually mean the ID (inside diamter) of your surface casing (or intermediate casing, or whatever) and the OD of the body of your drill-pipe (or kelly! Whichever is the smaller) not, in this case, the OD of your bit.

Another guide. As a rule of thumb, your bit OD shouldn't exceed two and a half times the drill-pipe OD (drill-pipe OD x 2.5) and shouldn't be less than twenty percent bigger than your biggest tool diameter (tool OD x 1.2) (that is your drill collar OD or, even, sometimes, the OD of an externally upset tool joint on the drill-pipe.

Figure 3-63A is an API list of oil field drill-pipe, not necessarily appropriate to water well drilling, but indicative of what we are talking about, because we are now going to talk about "upsets" (and there are plenty of those in drilling).

What does upset mean? Basically, if the bumped up (forged) part of the drill-pipe body on which the tool joint is fixed is bigger in diameter than its original OD to maintain a big hole in the drill-pipe/ tool joint throughout, then that is external upset. If you are looking for the drill-pipe to be flush on its outside diameter throughout the string say, for instance, to allow maximum uninterrupted passage of air flush (we know we have already said this but it is worth repeating) then, by necessity, the hole through the tool joint must be smaller, therefore it is internally upset.

You will gather that the actual meaning of upsetting is the thickening process whereby the thickness of the drill-pipe tube is increased at the ends only, either externally, internally or internally/externally (figure 3-64A) to allow more body for threading or for friction welding (or flash welding) your tool joints onto the body of the drill-pipe.

2) Tool Joint. What is a tool joint? It comprises the threaded sections on the ends of your drill-pipe.

Fig 3-63A

Dimensions and Weights

SIZE OUTSIDE DIAMETER		INSIDE DIAMETER		WALL THICKNESS		WEIGHT PER FOOT		WEIGHT PER METER	
						NOMINAL: THREADS AND COUPLINGS Lbs.	PLAIN END Lbs.	NOMINAL: THREADS AND COUPLINGS kg	PLAIN END kg
Inches	mm	Inches	mm	Inches	mm				
2⅜	60.3	1.995 1.815	50.7 46.1	.190 .280	4.83 7.11	4.85[2] 6.65	4.43 6.26	7.22 9.90	6.56 9.31
2⅞	73.0	2.441 2.151	62.0 54.6	.217 .362	5.51 9.19	6.85[2] 10.40	6.16 9.72	10.20 15.49	9.15 14.46
3½	88.9	2.992 2.764 2.602	76.0 70.2 66.1	.254 .368 .449	6.45 9.35 11.40	9.50 13.30 15.50	8.81 12.31 14.63	14.15 19.81 23.09	14.20 18.32 21.77
4	101.6	3.476 3.340 3.240	88.3 84.8 82.3	.262 .330 .380	6.65 8.38 9.65	11.85[1] 14.00 15.70[2]	10.46 12.93 14.69	17.65 20.85 23.38	15.56 19.29 21.86
4½	114.3	3.958 3.826 3.640	100.5 97.2 92.5	.271 .337 .430	6.88 8.56 10.92	13.75[1] 16.60 20.00	12.24 14.98 18.69	20.48 24.73 29.79	18.26 22.27 27.77
5	127.0	4.276 4.000	108.6 101.6	.362 .500	9.19 12.70	19.50 25.60	17.93 24.03	29.05 38.13	26.70 35.76
5½	139.7	4.778 4.670	121.4 118.6	.361 .415	9.17 10.54	21.90 24.70	19.81 22.54	32.62 36.79	29.43 33.57

[1] Tentative API Light Weight Drill Pipe for Grade E only
[2] Not API.

Material Strengths

PIPE GRADE	TENSILE YIELD STRENGTH OF MATERIAL		TORSIONAL YIELD STRENGTH OF MATERIAL		TENSILE STRENGTH OF MATERIAL		TORSIONAL STRENGTH OF MATERIAL	
	psi	kg/mm²	psi	kg/mm²	psi	kg/mm²	psi	kg/mm²
GRADE D	55,000	38.7	31,740	22.3	95,000	66.8	54,810	38.5
GRADE E	75,000	52.7	43,270	30.4	100,000	70.3	57,700	40.6
GRADE X	95,000	66.8	54,810	38.5	105,000	73.8	60,580	42.6
GRADE G	105,000	73.8	60,580	42.6	115,000	80.5	66,350	46.4
GRADE S	135,000	94.9	77,890	54.8	145,000	101.9	83,660	58.8

Theoretical Properties

Size O.D. Inches	Nominal Weight T&C Lbs/Ft	Tension Area Sq. Ins.	Polar Section Modulus Cu. Ins.	YIELD POINT IN TENSION					YIELD POINT IN TORSION*				
				Grade D Lbs.	Grade E Lbs.	Grade X Lbs.	Grade G Lbs.	Grade S Lbs.	Grade D Ft-Lbs	Grade E Ft-Lbs	Grade X Ft-Lbs	Grade G Ft-Lbs	Grade S Ft-Lbs
2⅜	4.85[1] 6.65	1.3042 1.8429	1.321 1.733	101,400	97,800 138,200	123,900 175,100	136,900 193,500		4,600	4,800 6,200	6,000 7,900	6,700 8,800	
2⅞	6.85[1] 10.40	1.8120 2.8579	2.241 3.204	157,200	135,900 214,300	172,100 271,500	190,300 300,100	385,800	8,500	8,100 11,600	10,200 14,600	11,300 16,200	20,800
3½	9.50 13.30 15.50	2.5902 3.6209 4.3037	3.923 5.144 5.847	199,200 236,700	194,300 271,600 322,800	246,100 344,000 408,900	272,000 380,200 451,900	349,700 488,800 581,000	13,600 15,500	14,200 18,600 21,100	17,900 23,500 26,700	19,800 26,000 29,500	25,500 33,400 38,000
4	11.85[2] 14.00 15.70[1]	3.0767 3.8048 4.3216	5.400 6.458 7.157	209,300 237,700	230,700 285,400 324,100	292,300 361,500 410,600	323,100 399,500 453,800	415,400 513,700 583,400	17,100 18,900	19,500 23,300 25,800	24,700 29,500 32,700	27,300 32,600 36,100	35,100 41,900 46,500
4½	13.75[2] 16.60 20.00	3.6004 4.4074 5.4981	7.184 8.543 10.232	242,400 302,400	270,000 330,600 412,400	342,000 418,700 522,300	378,000 462,800 577,300	486,100 595,000 742,200	22,600 27,100	25,900 30,800 36,900	32,800 39,000 46,700	36,300 43,100 51,700	46,600 55,500 66,400
5	19.50 25.60	5.2746 7.0686	11.415 14.491	290,100 388,770	395,600 530,100	501,100 671,500	553,800 742,200	712,100 954,300	30,200 38,300	41,200 52,300	52,100 66,200	57,600 73,200	74,100 94,100
5½	21.90 24.70	5.8282 6.6296	14.062 15.688	320,600 364,600	437,100 497,200	553,700 629,800	612,000 696,100	786,800 895,000	37,200 41,500	50,700 56,600	64,200 71,700	71,000 79,200	91,300 101,800

* Based on shear strength equal to 57.7% of minimum tensile yield strength.
[1] Not API.
[2] These sizes and weights are tentative and applicable to Grade E only.

A – Internally Upset   B – Externally Upset

Pipe dia.	Weight per ft.	Tool Joint	Joint dia.	Length		Length of upset (L)		Dia Upset Ext (D)/Int (D)
2⅜	6.65	2⅜ API-IF	3⅜	10'	20'	1¾	2¼	2.656/–
2⅞	10.40	2⅜ API-IF	3⅜/					
		2⅞ API-IF	4⅛	10'	20'	1¾	2¼	3.219/1.312
3½	9.50	2⅞ API-IF	4⅛/					
		3½ API-IF	4¾	10'	20'	1¾	2¼	3.824/1.937
3½	13.30	3½ API-IF	4¾	10'	20'	1¾	2¼	3.824/1.937
3½	15.50	3½ API-IF	4¾	10'	20'	1¾	2¼	3.824/1.937
4	14.00	4 API-FH	5½	10'	20'	1¾	2¼	4.500/3.340
4½	13.75	3½ API-IF	4¾/					
		4½ API-IF	6⅜	10'	20'	1¾	2¼	5.000/3.375
4½	16.60	4½ API-IF	6⅜	10'	20'	1¾	2¼	5.000/3.375
4½	20.00	4½ API-IF	6⅜	10'	20'	1¾	2¼	5.000/3.375

TOOL JOINTS ABOVE ARE DRILLQUIP'S RECOMMENDATIONS BUT WILL SUPPLY TO CUSTOMERS REQUIREMENTS
2⅜ IF–NC26   2⅞ IF–NC31   3½ IF–NC38   4½ IF–NC50

In many circumstances, usually related to ease of manufacture, drill-pipe is sold in random lengths and there are API standards for this. Range one is lengths of between 18' and 22', range two 27'–30' and range three 38'–45'.

A lot of "manufacturers" (note quotes) will say that proper upset drill-pipe (body upset) is not available and will offer you thick wall tubing onto which they will screw/friction weld your tool joints. As a very last resort this is OK but it makes them heavy and will restrict the capacity of your rig (and of your crew).

There is also "shrunk and welded", whereby the body of the pipe is heated and the tool joint, which has been machined near enough size for size with the tube, is pushed into the tube which cools around it, thus making a tight fit. The final operation is a ring of weld. Good for drilling with down-the-hole hammers where little torque is imposed, but for pure rotary work they can be a little weak.

How about screwed and welded? Well how about them. We don't like them unless they are prepared correctly, in which case they can become expensive. Let us explain.

They are made by threading the tool joint and tube with usually, a shallow thread, then running weld around the made up joint. The problem occurs if the "manufacturer" does not "torque up" the threads before welding. If they don't the thread will move when drilling torque is imposed,

then crack the weld and off comes your tool joint and you've got yourself a fishing job.

Proper API threading of tool joint to upset ended tube is excellent but in these days is becoming expensive because of high labour costs in making "all those threads", two for drilling and four for connecting per joint (length) of drill-pipe.

So we are left with flash or friction welding. Flash welding demands a plant of enormous cost and is a process of heating and pressure whereby the tool joint and pipe body are fused together without the addition of any other metal such as weld. The joint is immensely strong and used almost throughout the oil drilling industry.

Friction welding is nearer to our pocket because the plant to perform this operation is so much cheaper. Here the fusion is made by friction. That is turning and pressing — again, no additional metals. This is not only cheaper than threading but is marginally lighter so, a few more feet out of your rig.

So how do we summarise tool joint attachment? Well, it depends upon your operation. If you rotary and hammer drill then, if you can get it, go for friction welding, if not for screw on tool joints. If your work is purely down-the-hole hammer then drill tubes will suffice.

Now what about the tool joints themselves? Well, we've talked already about threads so, let's talk about choice.

Go for the biggest possible hole through the tool joint bearing in mind what you want your pipe to do. If you want it flush on the outside for air drilling then the hole will be smaller than "externally upset" (sometimes known as bottle joints) which you would want for an all rotary operation. As we are using a combination of rotary and hammer drilling we will have to compromise a lot but we would suggest you only consider internal flush or full hole thread forms, not regular for drill-pipe. This can be best illustrated by a case history.

We were asked by one of the charities to give a practical course on drilling and were given a brand new rig with mud pump, compressor, tools etc., etc., the drill-pipe being a peculiar animal called "HY" and in ten feet joints.

"HY" pipe was originally designed for shallow air drilling is three and a half inch OD and has two and seven eighth inch regular tool joints which have an inch and a quarter hole hole through them. Our mud pump was 5 × 6 therefore not overendowed with pressure (see later) — the charity specified the whole package without taking advice.

At one stage of the operation we were asked to drill a hole to three hundred feet using mud (Bentonite — horrible stuff, grisly details to follow), and long before we reached total depth the pump was clanking and banging fit to bust and blowing coal like you have never seen. The build up of friction (back pressure) through the tool joints as depth increased was just too much for the pump.

The stupidity of it all was that they could have specified two and three eighths IF tool joints in the same body size (three and a half inches OD) and got an inch and three quarter hole which would have been far better.

3) Length. We can't say much about the length of the pipe as this is governed by your rig, except to say make it as long as possible, because it saves time and weight. When you make a trip, the longer the pipe is the fewer joints you have to break-out, thus you save time. Also, two ten feet lengths of drill pipe are heavier than one twenty-feet length because you have one extra pair of tool joints (it's also cheaper because you save the cost of a pair of joints).

4) Grade. Now the quality of drill-pipe — grade. Never compromise here, never, never, never. Always go to the API specification and grade "D" minimum. There are equivalents in other national standards such as our British Standards, but make absolutely sure you get the correct thing. With your tool joints go for API standard specification or government authorised equivalents. Don't compromise yourself on any steel specification that goes down the hole.

Remember this. Drill collars tighten onto the thread shoulders whereas drill-pipe tightens onto the thread. In other words there should be a tiny gap between the shoulders of drill-pipe threads but collars, no.

So, what comes after drill-pipe as we go up? The rig? No, another sub, a wear-sub (or saver-sub) to protect your kelly/rotary head. Mind you, if you are using blow out preventers you would have to have a drill-pipe safety valve on top of your pipe, but that's another story for later.

Care of your Rotary String

We are going to say it again, thread protectors on everything, pipe, collars, subs etc., etc., — and for the reasons already gone into.

Use the correct grade of dope (grease) on your threads. Each country has its own range of dopes (in more ways than one — especially those who don't use thread protectors and thread dope) so take advice from the manufacturer of your tools and take that to your supplier of greases.

Always tighten your string together when going into the hole unless you want breakages. Good dope will make loosening simple.

Here is a little tip. Next time you buy new tools, before accepting them from the manufacturer, clean the threads of any grease, then run your fingers around them. If there is any sharpness or any lumps or pits do not accept them or it because, not only will they gall up, they will spoil other threads.

Now try the same test on your existing string — you'll get a surprise — discard as above and get a workshop to sort them out.

Discard all bent tools. What effect does bent pipe have on your hole? Hands up those who say you will get a bent hole (most people do, even some grey bearded pundits) — well you are wrong!

Your tools are going round and round, therefore the effect will be to increase bit gauge wear (a bigger hole? — yes), to wear the tools by rubbing against the side of the hole, and probably damage to the hole from rubbing and slapping. Get rid of bent tools — or have them straightened.

Drill with the weight of your drill collars leaving the drill-pipe in suspension, and never use pulldown.

Handle your tools with care. Do not drop them off the truck, you can bet your last rupee that they will bend if you do — you might even bend one of the crew.

If you are drilling pin down with "bottle" (external upset) tool joints, have the leading edge of the box end tool joint hard-faced, then the passage of fluid and cuttings won't wear it too quickly. Of course, if you are drilling pin up then it is the leading edge of your pin tool joint.

Keep a "pup" joint (short length) of drill-pipe in your arsenal. It will come in very handy we can assure you.

When holding your drill-pipe in the slips, yes slips are best, hold the pipe close to the tool joint to avoid any danger of bending from your tongs — of course you must use your tongs on the tool joint and not on the body of the pipe, otherwise you will damage it.

Why do we say slips are best? Because they are positive. As depth increases they become more and more positive. Hydraulic clamping is admirable — if you are drilling short holes. Ever heard of hydraulic failures causing tools to drop in the hole? We have. Ever heard of a helper being too quick on the laykey (slip plate) and losing the string? We have, and laykeys spreading open losing the string? We have. Keep your slips clean!

Make absolutely sure that your float valve (bit check valve) is working correctly every time it comes out of the hole and goes back in again. Stand it upright, and open and close the valve by putting a screwdriver or other suitable tool inside making sure that the spring does its job and that the valve moves freely — if not, repair it.

Then turn the valve the other way up and fill the cavity with water, if it leaks, it's no good and needs repairing.

Auger strings

We won't dwell too long here because augers are rare in "our" business so let us list some do's and don'ts:-

Do find a safe way of handling because they are awkward and heavy — perhaps the way we handle drill collars.

Don't use any retaining pin that might be even a little soft because auger torque requirement is enormous and will shear a soft pin when the hexagon connection starts to wear. Book Nine gives a formula for this calculation, try it, and you will see.

Do have your bit about half an inch bigger than your flight in diameter when drilling vertically.

Do make sure the construction is good, with nice heavy plate, substantial centre and good welding. A guide to pitch is "eighty percent". So, what is pitch and what is "eighty percent"? Pitch is the distance between the flights related to diameter. "Eighty percent" means that, if you have a ten inch OD auger then the distance between the flights would be eight inches — QED.

Do use a good guide to centre them when starting the hole, they can be a bitch if they go off. Talking of bitches:-

Do ensure that your bitch is in good condition and the correct size. What is a bitch? Its like a slip plate that holds the auger string in the working table (only top drive rigs please) when making up and breaking out.

Don't bend them, they can be a devil to straighten.

Cable percussion string

Simplicity itself. If you follow the rules — and safety is the main rule both for the crew and for the tools as these components are heavy. But let's look at the tools first. Figure 3-67A illustrates a typical string and, having covered bits already, we will start with:-

The drill stem. This is your string of drill collars. They are solid (no hole down the middle) and they give weight to your bit for drilling. The bigger the bit the more the weight, the harder the formations the more the weight etc., but only

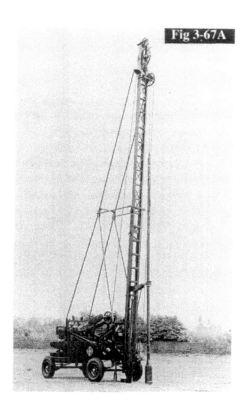

Fig 3-67A

and fishing, but that comes later.

Jars come in different strokes according to the rig and the driller's needs. Check them for wear, especially on the "clanging" end of the links. They can be very dangerous if they break.

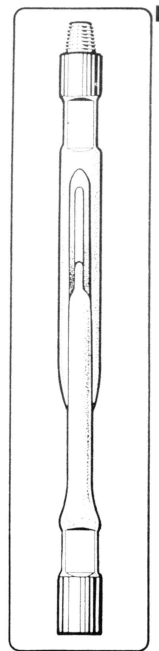

Fig 3-67B

within the capacity of your rig. Another point to remember — like drill collars, you need to put your weight over a short length — long lengths can be unwieldy — so, use the biggest permissible diameter stem.

By the way, we don't use cable percussion in hard formations do we!

When calculating the weight of tools in the hole don't forget the weight of the wire-line — that is down there too!

Check your drill stem for damage to threads or bending every day. A bent stem can get you into all sorts of trouble — just imagine what it looks like in a hole and the trouble it can cause you. Think about it.

Jars. As we are going to use our rig in soft formations only, then you will always use jars, (not really necessary in hard formations). What are jars? Have a look at figure 3-67B then imagine a nice sharp bit thumping its way into a sticky lump of something in the hole. The walking beam opens the jars with a clang (the two links coming together) jerking the bit out of trouble — that's what a jar is. Jars are also useful in casing

Rope socket. Attaches the rope to the jars. Make a good job of the joint (holding of the rope in the socket) — we don't need to say any more about that do we? — "a good fisherman is not necessarily a good driller". Make sure the rope mandrill is free to move inside the rope socket (see previous comments in Book One).

Bailers. You know what they do — we told you in book one — OK you've forgotten so we will say it again.

When drilling you make cuttings in the hole and, unlike rotary (hammer as well) drilling, you can't pump them out, so what do you do?

In a dry hole you pour some water into the hole after you have withdrawn the bit (wet hole has its own water — hasn't it?) then, on the bailing winch, you run the bailer down. Figure 3-68A shows alternative types of bailer but they all have one thing in common — a flap valve of sorts on the bottom. So, the valve opens when it encounters the mixture of water and cuttings which is taken into the bailer, the valve closes on the mixture, and it is lifted to surface for disposal. This takes a number of cycles.

Like all tools, the bailer has to be looked after. Check that the valve works and the bail (the loop at the top for attaching to your wire-line) is not broken or wearing out.

About the Well-Site

A few tips on keeping the well-site tidy, with special reference to the drill string.

Keep your drill-pipe off the ground, preferably on trestles, then you not only keep it clean, you have access all around it for things like greasing and measuring (some pipe comes in random lengths don't forget). Put nice big stops on the end of the trestles to stop the pipe rolling off, and don't let people walk on pipe — ankles can break if two come together.

Make sure the trestles are set up so that you have good access to the rig when making up or breaking out pipe.

Have at least two lifting plugs (see earlier) available for each size of thread in your string. When you trip on the winch one is working and the other waiting — this saves an enormous amount of time.

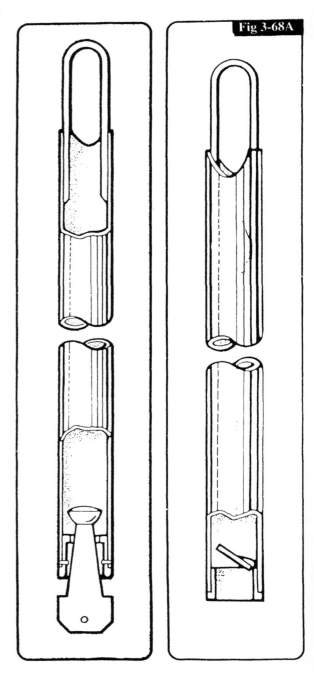

Fig 3-68A

Another point or two about lifting plugs. Ensure they are of the swivel type and that they are up to capacity (of your string) and that the bearing is well greased. If not, wire-line spin and bad "tonging" can undo them and down go the tools. Your lifting plug should also be of the breathing type, meaning that it has a hole through it, thereby largely overcoming suction in a wet hole — known as swabbing — which can "pull in" soft formations.

And what about your beautiful drill collars? They have to be safely stacked and above the ground and the greatest care taken in loading and unloading not only in and out of the rig but on and off your support truck. A story about mis-handling of drill collars:-

A crew were setting up to drill a well in a desert and the driller was kneeling on the sand attending to something or other. A helper pushed a twenty foot, six and a quarter inch drill collar off the truck and it crashed down onto the back of the drillers legs. One leg was in soft sand and was only bruised, the other was on a patch of hard sand — that driller has never walked normally since — his lower leg was smashed.

Again, put your drill collars where they will be most effective for handling. Put a "deadman anchor" away from the rig with a pulley, this way you can run a rope around the pulley to the free end of the collar when winching and ease it along. Don't, whatever you do, allow anybody to be under the drill collar when it is being handled.

You, the driller, should have sight of all operations such as tool handling, pumps etc.

Keep crowds (people) well away from the rig — accidents can happen.

If there is a prevailing wind, set the rig up so that the wind takes exhaust gases away from the crew and if you have blow-out preventers make sure you do the same thing with the blooey.

Another thing. Some drillers like to have a little track and trolley for tool handling. This is simply a small wheeled thing into which is put the free end of the drill-pipe (or collar), and as the pipe is "winched" up the trolley, runs along a simple track, and the other way around when winching down.

Keep all vehicles, fuel supplies, mud sacks etc., well away from the rig, except heavy mud (if you need it) for weighting to overcome abnormal formation pressures. That must always be handy to the rig.

If all these things are done automatically everyone is free to do their job safely and efficiently.

4 Flushing Systems

Foreword to Book Four

"If you cannot clean the hole don't start".

Many of you will know about this because you have got yourself in and out of trouble in your "youth", and will know what it is like to feel your heart thumping away in the chest waiting for cuttings to come out of the hole, or watching the torque build up on the gauge or trying to free stuck pipe, or just taking many times longer to drill the hole than necessary and yet the rules are relatively simple. Now.....

Read on.

Flushing Systems

Here we will deal with the three basic flushing systems most widely used in water-well drilling today, namely: air, foam/polymer injection and mud. We will touch later on reverse circulation which is now less widely used internationally.

Please note the use of the word "FLUSHING" and not circulating. Mud and water are circulating systems, compressed air and foam/polymer are not.

The cleaning of the hole is the most important single factor in drilling, as an inadequate system which does not clean the hole will cause the greatest of problems. No attempt should be made to even start drilling unless you are satisfied that firstly, the system is correct for the conditions to be drilled and secondly, that there is sufficient pressure and volume in the "pump" (includes compressor) to maintain at the very least minimum annular velocity.

Let us just dwell on these points. Book Nine gives a series of formulae which will tell you whether your rig can do the work in hand and if your "pump" can clean the hole. These formulae are quite simple to use and should always be applied to every job you tender for and, just as importantly, when you think about buying a new piece of equipment — even the holes through drill-pipe etc.

But of course, if your rig was set up correctly *from the beginning* with the parameters of future work taken into consideration, then all should be fine. As already said, the formulae in Book Nine should be used when going into the purchase of equipment. Don't let poorly qualified manufacturers bamboozle you into buying the "all singing, all dancing — flavour of the month — pretty pretty" rig they just happen to have in stock. I have yet to see an efficient piece of drilling equipment that didn't look anything but functional.

Let us illustrate the above with a story about something that happened just recently. We were commissioned to do some drilling tests (the formulae were used to judge the adequacy of the unit) and amongst the equipment was a brand new compressor supplied by a very famous manufacturer, along with which came a *top* service engineer. When asked why he was running the compressor with doors open (always check this point with your supplier as almost always you should run with the doors closed), he said "... the fan sucks through the radiators therefore the doors must be open...".

We took him to his compressor, shut the doors, and instead of "sucking" the fan was blowing through the radiators, in so doing drawing air across the engine and compressor. Therefore the doors had to be kept closed *as written in his company instruction book*.

Back to business. In previous lessons we have explained how best to set up a rig based on work forseen and using the formulae so we won't take up valuable time here.

A few rules about placement of "pumps": we are putting pumps in inverted commas because they include compressor, foam/polymer pump, and mud pump. After all, a compressor pumps air doesn't it?

"Pumps" should always be visible to the driller who, wherever possible, should have control over their function. This latter point can be difficult where a mud pump is involved (though not impossible) but the driller should still have a clear view of the pump and *pumpman*. A lesson here.

An inexperienced operator is at the controls and is tight in the hole. He is getting no "returns" and doesn't know that the relief valve has blown on the pump because he can't see it — result, panic.

Mind you, there is something else to be said here. What is written in all these books is based on both theory and practice and any driller who has never been stuck in a hole, or never experienced other such numbscullian, heart jerking situations, should be sanctified because his guidance comes from someone greater than us: in other words we have been through it.

Enough said about that, so let us proceed with systems.

Compressed air flushing

Remember these three rules before you even start up:-
1) Compressed air is dangerous. Always make sure that you have safety devices (chains, cables etc.) on *both ends* of the delivery hoses *and* that

the strength of the hose *and* its couplings is sufficient to operate at the delivery pressure of the compressor.

2) Compressed air is the most expensive form of power that you will ever come across in the drilling industry and probably elsewhere, so don't waste it. Get those leaks patched up in a hurry, don't run the compressor when not needed and don't use excessive amounts of air when rotary drilling — you have the formulae, use them.

3) Compressed air drilling seems so easy. You just plug in the compressor. Start up and away you go. No pits to dig, no water to get, etc. etc. — don't let it fool you. In the right conditions it is as easy as that, but out of its effective environment it can be positively harmful.

Now, how about that hose you have connecting your compressor to your rig which is safely attached at both ends by safety devices? It has the correct pressure rating and the couplings are wonderful, but is it the correct diameter to deliver the air you require and, just as important, are those fittings between the hose couplings and the rig/compressor of sufficient diameter and pressure rating?

A reasonable hose/couplings/fittings diameter is two inches inside diameter all the way through — minimum — and if you are operating a remote compressor keep the length down to about twenty metres to save on possible pressure losses. Remember, 1 psi is equal to 2.31 feet head of water, so for every extra psi you get down the hole you can get that much more drilling under a head of water.

The size of the compressor will normally be governed by the consumption of the down-the-hole hammer, unless you are rotary drilling with air only, which is a little bit "old fashioned" these days with the advent of foam/polymer additives. We like to see the possible volume of air from the compressor about fifty percent more than the hammer requires *when drilling* with a blank choke, not when blowing. This extra percentage makes allowance for the extra air for hole cleaning if additives are not in use and allows the compressor engine to work at a reasonable level.

Another story which might illustrate much more effectively remarks made in the previous paragraph about chokes, volumes of air and the like.

We were asked to look at the possibility of drilling with a nominal six inch (bit) hammer on 3½" pipe inside 8" casing with a 425 cfm, 170 psi compressor.

With a blank choke this particular hammer required 325 cfm at the 170 psi available (remember, the higher the pressure the higher the volume) and would have been comfortable at that, but if you calculate the annular velocity (minimum 3000 fpm) through 8" casing on 3½" pipe you would see that we would have been very short of cleaning power: by injecting a foam/polymer mix the problem was overcome and the hole completed in good time.

"But", said the drilling supervisor, "why use foam when there was plenty of groundwater in the hole which would have done the same thing as the foam mix? Everybody knows that, look at it coming up the hole."

Our reply? Everybody does not know that because what you are saying is, with all due respect, incorrect. Groundwater is a hindrance to hole cleaning and the more the water the greater the problem.

You see, the column of water is hanging on a cushion of air above the bit which slows down the exit of cuttings from the bit (called chip hold-down), which are ground and re-ground until the majority become dust and they "queue up" for their turn to go up into the water and out of the hole. This is exactly what blocks hammers if the check-valve is less than efficient. Imagine a cloud of cuttings just above the bit above which is a mass of water. The hammer is shut off, say for pipe change, the check-valve is a bit slow and the mass of water forces the dust into the hammer — you've got a trip on your hands.

How to prove this point? Two ways:-

Next time you are in this situation and you are about to change the drill-pipe, take samples of the groundwater coming up the hole and see how much solids are in there, then inject a bunch of foam and check again, you will see what was left down there.

The second way is to place some receptacle for catching samples at the top of the hole and drill two or three feet without foam mix, then pump foam for a foot or two and shut off the foam for the rest of the pipe. In the receptacle you will find a beautiful sandwich of chippings *within two* outer layers of dust.

These are proven facts.

We want to go over chokes again, because understanding their operation is very important. They are, of course, used mainly in down-the-hole hammers but can be adapted to air rotary. For the purpose of this discussion we will, again, concentrate on the hammer.

The air that operates a hammer cleans the hole on discharge from the bit — or it is supposed to. Sometimes, maybe because of oversized bits or drilling inside large casing, there isn't enough to give the required minimum annular velocity. Most hammers these days have a method (usually a central tube or the like) of passing extra air to make up the deficiency, and you control this quantity of air by fitting chokes.

The choke is usually a sort of rubberised plug which can be just a plug (blank choke) or can have a hole drilled through it. A set of chokes will comprise a blank, and a number of others each with a different sized hole. The larger the hole, the larger quantity of air you can pass through it, which will supplement the air used to operate the hammer. You use the one which will give the air you need.

There is one manufacturer of hammers who uses a plastic cylindrical choke which is horizontal in the check valve and through which the air is passed across "flats" of varying dimensions (according to need) cut on the side.

Just in passing. Large hammers on relatively small drill-pipe which would normally require vast quantities of air to clean the hole can be drilled using the blank choke, assuming you have a compressor big enough to operate the hammer efficiently, and foam injection. A job we did in Hong Kong was a 21½" hammer on 5" drill-pipe with foam — excellent results.

When to use "pure" air flushing

The best answer to that question is found in when *not* to use it, and that is very simple — never in any unstable formation or any formation that is likely to be eroded due to high pressure and velocity at the bit, unless that is, you use additives.

It is rare for formations to collapse on their own. Almost all hole collapse is induced by using the *wrong* flushing system.

Again, we can best illustrate this by a case history — why not? We learnt from such experiences and so will you.

We had completed a major well with some success and, as we were "experimenting" with new systems, we were asked to pass the information on the work done to another crew (and rig) so that they could carry out a further programme of wells in that area.

We did this, saying that there were clays, soft sands, laterite and further sands in ascending degrees of hardness. We also told them the drilling system used — rotary (drag bit) with foam/polymer flushing onto bedrock, installation of casing, and through with a down-the-hole hammer to T.D.; yes, says the engineer.

Three weeks later he abandoned the well (the first) due to collapse of the hole, temporary loss of the casing, the top of the hole opening up (thus the rig falling in — *twice*), and an inability to maintain the hole *in any way*.

He had put a down-the-hole hammer on from surface and the force of the compressed air had blown the sands away, thus cavitating the hole and inducing massive collapse. From the word go he was on his way to disaster, and remember:-

ONCE A PROBLEM STARTS IT IS DIFFICULT TO STOP — IT IS BEST TO START OFF WITH MAXIMUM SECURITY.

Foam/polymer injection flushing

Since the end of the 1939/1945 war, after things got back to normal, there have been three major advances in drilling: the top drive drilling rig, the down-the-hole hammer (commercially viable) and foam/polymer injection (flushing) systems.

That's how important foam/polymer systems are.

Notice how this system has been written "foam/polymer". That is because it is a system comprising a number of ingredients, foam being one of them. With the foam you use one or more polymers to overcome various problems, water to mix it in and compressed air to power the mixture. Let us start this most important section by going a little more deeply into what the system does and how it overcomes some of the problems you are likely to encounter.

Foam lifts cuttings at any speed you fancy. You can appreciate this more by imagining a vertical (or angled) conveyor belt with buckets attached that is lifting gravel. You can vary its speed as you wish according to your needs and the gravel will not slip backwards because the buckets have got hold of it. That is the important thing to remember — *it will not slip back even if you stop the conveyor belt — and it is the same with a good, well set up, foam/polymer system.*

Foam will fill cracks in the ground, cavities and great voids in its own sweet time and when those troublesome voids etc. are full, then the foam will continue its relentless advance upwards. Even if you stop it, it will remain in control of the debris (cuttings) awaiting your command for it to reach surface, deposit its burden and then conveniently expire in a very short time, thus making way for what is following.

Put your thinking caps on and work out what has been said in the previous couple of paragraphs about overcoming problems, and what problems — we will come back to it later.

There are many different kinds of polymers but they are divided into two types, natural and synthetic. We do not recommend the use of some natural polymers as they can induce bacteria which cannot be wholly cleaned out after the well is finished, thus causing fungus build up in the hole which, whilst not necessarily dangerous to health, can block off screens etc.

Good foams will mix well in fresh or brackish water but their carrying capacity is reduced when groundwater is encountered, due to dilution, so you need a polymer to tighten the bubbles of foam preventing dilution. That is one polymer.

You have to build up viscosity in the mix to "slick up" the operation of the system and give it its carrying capacity — thats another.

You've got to prevent groundwater hydrating (making wet) difficult formations (sloughing shales for instance) thus causing collapse — thats another.

When you've got all those you have to put something else in to make them all mix easily without causing great blobs of polymer (fish-eyes) forming, which are useless for drilling and can block your pump, that's another. Oh yes, you

need an injection pump as well.

Complicated isn't it? No it isn't, because all these items can be mixed by the manufacturer to your needs in one package. Thus the driller can take a pinch of this and a drop of that, put it all in water and providence (and the system) will do the rest.

Find yourself a good manufacturer who has drilling experience, tell them your problems, and they will come up with the goods. They will also have available a "general" mix of polymers which you can use as a trial.

Your foam pump should be capable of injecting the required amount of slurry (mixture) and it is rare that you will ever have to inject more than two gallons a minute so, rather than looking at the maximum delivery from the pump, you look at how low down you can control its operation. Generally speaking a pump that is controllable from 0-5 gallons per minute will suffice. Discharge pressure from the pump should be at least twenty percent higher than compressed air pressure in the rest of the system otherwise the slurry will *not enter the airstream.*

Figure 4-75A is a diagram of a typical foam/polymer injection circuit and perhaps it is better if we run through it point by point:–

The suction hose, which should be of clear plastic (see later) is put into your tank full of mix (foam/polymer), the compressor is connected to air inlet (a) and discharges from the circuit (b) to your standpipe (the rigid mud/air line which runs up your mast to which your mud hose is connected, not a short bit of casing stuck in the top of a hole — that is a conductor) or even direct to your swivel.

Compressed air passes through valve (c), which is the master valve, some of which is then diverted to another valve (d) which controls the speed of (therefore delivery from) the pump if it is air powered. The balance will go on to another valve (e) which controls the amount of air to be injected into the hole *with the foam/polymer mix.*

When valve (d) is operated the pump draws mix from the tank through the suction hose and passes it through a non return valve (f) into the stand-pipe via a venturi tube (g). Air that has not operated the pump passes valve (e) and goes through another non return valve (h) where it meets the mix coming out of the venturi tube (g).

There, because of the venturi effect, foam is made.

It is important to remember that foam should be made at the rig and not down the hole otherwise you won't get full efficiency from the system. The foam is then passed down the hole, cleans the bit, cools it etc. and goes up the annulus to the surface driven by the air from valve (e).

Non return valves (check valves) (f) and (h) are an essential part of the system. The non-return valve on the air side prevents foam entering the compressor and the other prevents compressed air going into the foam pump.

Foam/polymer drilling is visual. This means that once valves (d) (quantity of mix) and (e) (quantity of air) have been adjusted to give the correct consistency of foam (smooth and thick like shaving cream), those valves can be disregarded when shutting the pump on and off when say, changing drill-pipe, the master valve (c) will shut off the circuit and open it again leaving other valve settings in place.

If your pump is hydraulically driven, make sure you have complete control of pump speed otherwise you could use too much (or too little) mixture.

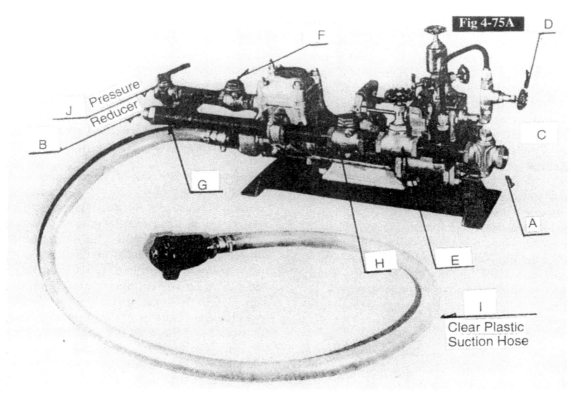

A. Air inlet

B. Foam to standpipe

C. Master valve

D. Pump control valve

E. Air control valve

F. Non return valve

G. Venturi system

H. Non return valve

I. Clear plastic suction hose

J. Pressure reducing valve

When to use foam/polymer injection

You could say it should be used in just about any situation except where hydrostatic head is required to keep the hole open, such as water bearing running sands. But that would not be the right thing to do, especially as you are now required to answer the question asked earlier in this book, i.e. to anticipate a number of situations from given clues.

1) *Hole cleaning*
Foam/polymer injection will clean the hole efficiently, irrespective of annular velocity in almost all conditions. A great deal has already been said on this point so we will move on — but it is a fact.

2) *Lost circulation*
A story. Again we were asked to supervise the drilling of some large diameter wells in the Homs area of northern Libya and, on arrival in Tripoli, an experienced drilling engineer went along to buy the bits from a major manufacturer. The conversation went like this:-

Him — "Where are you going to drill?"
Engineer — "Homs"
Him — (with a shocked expression) "But that area is undrillable"
Engineer — "Why?"
Him — "Because it is dolomitic limestone and the cavities can be as big as this room and just about everybody has failed because of total loss of circulation. There is more pipe stuck in the ground around there than you can imagine."
Engineer — "But we are using foam/polymer injection."

He didn't answer, just took the money and handed the bits to the engineer, although a wry smile was noticed on departure.

In the first hole we encountered cavities as much as 10' in depth but the hole was completed very quickly (already described in a previous lesson) and many more followed — we had no failures whatsoever.

In lost circulation zones where you have no returns, watch your clear plastic suction hose (remember?) to make sure that the mix is going into the hole. Listen to your compressor to make sure air is following and watch the torque gauge on your rig to ensure that your bit is not getting tight; if it is, ease off, give it a good clean, then continue.

3) *Water (mud) losses*
Where the formations absorb mud (this is not lost circulation). It is impossible for foam/polymer to be lost in this way.

4) *Spurt loss*
Where water separates from the the mud and enters the formations. Again impossible with foam/polymer as there is no free water in the foam/polymer and when it is carrying ground water, which it does with consumate ease, the groundwater cannot be released into the formations due to the strength of the bubble. Separation takes place on surface. That, incidentally, is how you know when you have groundwater — foam/polymer drilling, properly set up and recognised, helps in ways we haven't even thought about yet.

5) *Formation control*
Before we go on here let us say again, foam/polymer will not be effective in conditions where hydrostatic pressure is needed to hold the hole open — say, water bearing running sands for instance.

The velocity (or lack of) of the foam going up the hole is all important if you are in soft conditions because the faster it goes the more air you are using, therefore the greater pressure and velocity you have *at the bit*. This is one of the most important lessons to be learned. What is crucial is the behaviour of your fluids (all of them) *at* the bit because it is there that the formations are subjected to the greatest stresses.

As the fluid exits from the bit, it will blast onto the bottom of the hole first and then all around the bit blowing your soft formations away and causing massive erosion. A cavity is created, down goes your annular velocity, debris remains around the bit, the formation collapses and you are in big trouble — the problem has started and you can't stop it. You can't put the formation back can you? All that's left is to install casing, and you haven't finished with the soft formations yet, or cementing. What a waste of time and money.

THE SOFTER THE DRILLING THE MORE DIFFICULT IT IS — THE HARDER IT IS THE LESS DIFFICULT.

Now, if you had a system that would clean the hole efficiently, impose no pressure on your poor old formations and would not erode the hole on the way up due to its low velocity, life would be pretty good wouldn't it? Well you've got it and its called "low velocity foam/polymer injection".

"Sir" said the driller "we are drilling hard rock and it is blue".

He was reporting his progress for the day, which was very poor.

"Rock in this area?" says the engineer — this was a desert region and the hardest we had ever had before, after many many wells, was some weakly cemented sandstone. These were 17½" holes to an average of 500' and we were using foam/polymer for hole cleaning and a soft formation rockbit (the client's choice).

The engineer went along to investigate and sure enough the foam coming out of the hole was bright blue in colour (another good point about foam/polymer drilling — it changes colour in different formations and you *can see* what you are drilling) and drilling.was slow.

The engineer took a sample to the geologist who immediately recognised it as the clay from which bentonite (montmorillonite) is produced, and if we had been using water or basic mud that formation would have turned to a slurry on being contaminated (hydrated) with water and even later groundwater had no effect on it.

Our polymers, mixed in the foam, completely controlled that very difficult formation.

This case history only goes to show what can be done with this extremely versatile system and, when it comes to mud, what enormous benefits there are in using polymers — the possibilities are endless.

FOOTNOTE: That client never did use drag bits when he should have done because he was completely sold on oilfield practices; there are some strange people about.

The main usage of high velocity foam is in down-the-hole hammers but even that can vary depending upon the annulus. If you are using a "jumbo" hammer on small pipe then the velocity will be slower. Much has already been said on this so we will pass on.

Sometimes you might want to compromise on velocity (now we are really getting technical). Say, for instance, if you have to use a core bit with the wrong sized (too small) ports, then you can turn up the air slightly to avoid blockage of the bit which would occur in an ultra low velocity injection rate in softer formations.

You must avoid the air breaking through the foam, meaning the air must lift the foam and any view of extraneous air coming up the hole without any foam in it means you are breaking through. What to do? If permissible, turn the air down, if not (small ports in the core bit — but not too small) pop a bit more polymer in the mix.

Once you get used to the idea of the vast potential of the system then the choice of mixes, velocities etc., is yours — use it, you will be rewarded by much greater drilling efficiency, therefore production.

So how many did you get? All of them? That's good because what has been said is only a fraction of what you can do with the foam/polymer system which means, like us, you don't know it all, and the thirst of knowledge of an active mind, like yours, is a great gift.

The actual amounts of foams and polymers needing to be mixed with water to get an efficient system will vary according to the manufacturer and according to the work being done, so take advice from someone "who knows" — there are plenty of "cowboys" about, so your first question to them should be "...how much actual drilling have you done with the system (or indeed any other system), and I mean 'hands-on experience' in the field...". That will sort out the sheep from the lambs; and don't be frightened to take advice from more than one person.

Always have two suction tanks at the rig because you will be drawing (sucking) slurry (mix) from the first and the second, if not being mixed, will need a little time for the polymers to completely hydrate and reach maximum yield (viscosity). This time will vary according to the polymer but should not exceed more than fifteen to twenty minutes — never let them both run dry because, as sure as eggs are eggs, you will have a problem in the hole and nothing with which to flush the problem away.

Make sure your foot valve and strainer on the suction hose are in good condition and that the strainer will filter out all those ghastly bits of cuttings that your drill helper has just washed off in your nice clean slurry. In other words keep your tanks clean, because your poor old pump will suffer no end from trying to pump solids.

Right, now that you are an expert at probably the most effective yet simple (and it is simple to use — you will see) form of flushing, we will go on and talk about something else.

Reverse circulation

This is where, instead of your flushing fluid going down the drill-pipe and up the annulus between the side of the hole and the outside of the drill-pipe, it goes down the annulus and up the inside of the drill-pipe. Everything is in reverse, hence reverse circulation.

If the speed of the downward movement of the fluid (down the annulus) is controlled (maximum 60 feet per minute) then hydrostatic pressure is exerted on the walls of the hole, thus holding up the formations — too fast and the hole will be eroded.

This system was first thought up in Holland which, as you know, is the delta of the river Rhine, hence almost the whole country is made up of soft silts and sands. To illustrate this, the railway station in Amsterdam is built on 12,000 piles.

Bangladesh is the delta of the river Ganges, therefore is soft, therefore reverse circulation drilling is predominant. Other countries like this are Burma (the south), Egypt (Nile delta) etc. etc.

In addition to controlling the maximum speed of the fluid (usually water or a light mud) down the annulus you have to have a normal sort of velocity up the inside of the pipe to clean the cuttings from around the bit. That speed is around 120 feet per minute, which is a good average for direct (normal) circulation with water.

Therefore you will see that, to use this system, you have to be aware that the drill-pipe diameter (inside) and the diameter of the hole are critically interrelated and that reverse circulation should only be considered in the drilling of quite large diameter holes and, from our point of view, nothing less than fifteen inches.

Drill-pipe is usually flanged and these flanges are quite large in diameter. On six inch pipe for instance, they are around eleven inches (figure 4-80A). Therefore you have to drill a big hole anyway. Screwed pipe is available but is very expensive.

Figure 4-79A shows a typical layout of a reverse circulation site with a mud pit connected to the hole by a lined trench along which the water runs from the pit into the hole. *The hole must always be kept full of water/mud or it will collapse.* Discharge from the drill-pipe should either be run over a shale shaker (see later) to get rid of the cuttings or along a series of channels — as with normal water/mud circulation — and into the pits.

Ample "top-up" water must always be available in case of major losses into the formation which will result in the lowering of hydrostatic pressure and cause collapse.

From the bit through to the exit from the discharge hose (drill-pipe, swivel, discharge hose) the inside diameter should be the same eg. six inches to *avoid bridging the cuttings*. If the movement inside stops, you can break suction, then you are in a mess because, not only is your hole full of cuttings but so is the drill-pipe. Then you really have problems.

There will be times when you get large cobbles or boulders around the bit which either won't go up the pipe because they are too big or they are too hard for the bit (drag bit — remember?) to break up. At such times it is a good idea to have some sort of a grab e.g. orange peel, that you can run in the hole (after removing the drill-pipe of course) on your sandline winch from time to time to remove these obstacles.

There are two ways of getting the water/cuttings to come up the drill-pipe, suction, and air injection.

Suction lift

Before we start, remember your physics and that the maximum suction "head" is 27 feet. So to be on the safe side, and to make allowances for the

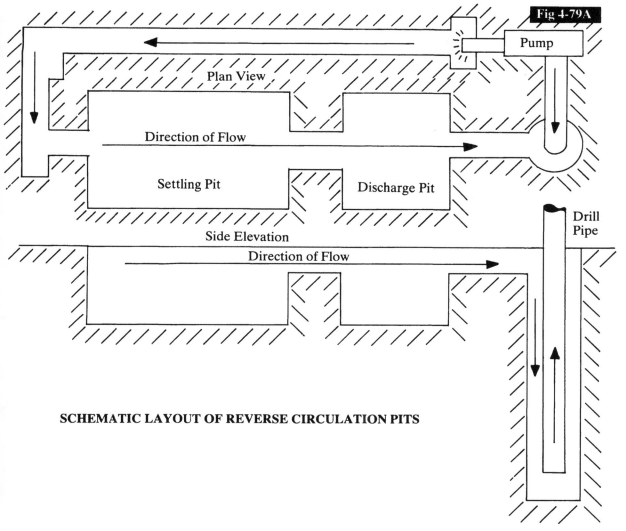

Fig 4-79A

Pump

Plan View

Direction of Flow

Settling Pit

Discharge Pit

Drill Pipe

Side Elevation

Direction of Flow

SCHEMATIC LAYOUT OF REVERSE CIRCULATION PITS

distance from ground level to your working/rotary table and running the drill-pipe up and down the mast when hole cleaning etc., your drill-pipe should be in (about) 10 feet joints and not much more.

For suction lift you need a centrifugal pump with an inlet matching the inside diameter of the discharge (again) with an inbuilt vacuum pump or system. The pump should be for "solids service" which means it must be able to pump sometimes quite large solids.

Between the exit from the discharge hose and the pump inlet it is better to have a "stone trap" to take the larger cuttings which should be cleaned out every time you change the drill-pipe. What is a stone trap?

A stone trap is a largish vacuum sealed container fitted into the system into which stones will drop — simple enough isn't it? But make sure the connections are vacuum sealed also.

Now we have it all together, how does it work?

With the bit in the hole which is full of water and with an ample supply of water in the pit and the discharge hose connected to the inlet side (remember that, not the discharge side) of the pump with the stone trap in between, start the pump and vacuum system. A vacuum will be created inside the air tight drill-pipe and swivel etc., and gravity will push down on the water in the hole forcing it into the vacuum in the system coming out from the discharge hose, into the water/mud pits thence into the hole again and you have *reverse circulation*; now drill on. Wonderful isn't it, and what a great mind thought that up?

Figure 4-80A shows a typical flanged type drill-pipe for reverse circulation but illustrates that for air lift (see later); just ignore the air pipes (small) running down the sides. Great regard should be paid to the sealing of the joints so as not to upset the vacuum; always use a gasket between the flanges.

A word of warning here. Because of frictional losses inside the drill-pipe that could break the vacuum we have found that a good maximum drill depth with the suction lift system is about 400 feet. Going deeper? Then use air lift.

Air lift reverse circulation

Before we start, this system is inefficient for the first fifteen feet (see theory of submergence) or so at the top of the hole so you should have some sort of vacuum system standing by. But, it has a much greater depth capacity than the vacuum.

The theory of reverse circulation in this function is the same as with suction lift except that, instead of creating a vacuum inside the drill-pipe and letting gravity do the work we are going to inject compressed air at a given point (or points) into the drill-pipe. This makes the water lighter (less dense) than the water below, which will then rise to take over the lighter area which is then aerated and moves up and so on and so on — we now have air lift reverse circulation. Drill on.

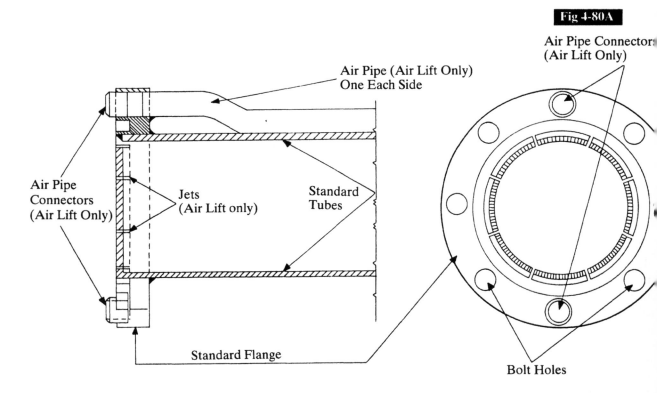

Fig 4-80A

COMPOSITE SKETCH OF REVERSE CIRCULATION DRILL PIPE

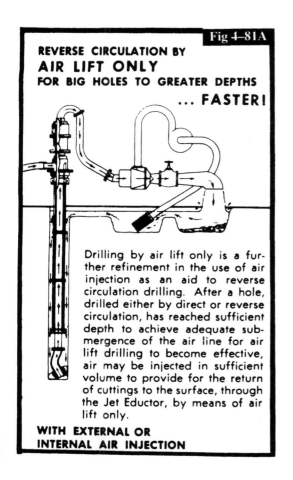

Fig 4-81A

REVERSE CIRCULATION BY AIR LIFT ONLY

FOR BIG HOLES TO GREATER DEPTHS

... FASTER!

Drilling by air lift only is a further refinement in the use of air injection as an aid to reverse circulation drilling. After a hole, drilled either by direct or reverse circulation, has reached sufficient depth to achieve adequate submergence of the air line for air lift drilling to become effective, air may be injected in sufficient volume to provide for the return of cuttings to the surface, through the Jet Eductor, by means of air lift only.

WITH EXTERNAL OR INTERNAL AIR INJECTION

What changes are there from suction lift? They are:-

In the area of swivel you have to have an air inlet system but still maintaining the inside diameter of the swivel (remember? to avoid bridging of cuttings). This compressed air is passed down the drill-pipe (outside the main discharge area — see figure 4-80A for schematic view) to a drill-pipe which has channels machined into it and which is known as a "jet pipe", and the compressed air is passed through these channels (jets) into the inside of the drill-pipe, which is full of water, which becomes aerated and the cycle begins.

We like to see a jet pipe every seventh pipe (for 10 feet joints of pipe), because as you go deeper so a pressure is built up by the water in the hole which eventually cancels out the air pressure —

remember, 2.31 feet head of water is equal to one pound per square inch.

You don't, of course, need a centrifugal pump/vacuum system (except for the top of the hole) nor do you need a stone trap because the discharge hose is open to the shale shaker or channels, but you do need a compressor.

The air compressor need only be quite low in volume because all you are doing is putting a relatively small amount of compressed air into the system, but pressure should be looked at carefully. The higher the pressure the fewer jet pipes you need. But don't go overboard here — keep the whole thing simple and uncluttered.

An air compressor of about 350cfm capacity at a pressure of about 150 psi will really drill some very big holes.

Jet (eductor) reverse circulation

This is another sort of reverse circulation drilling which is similar in overall principle to the suction lift system except that a jet pump is used in place of the suction pump, thus avoiding cuttings passing through the pump (figure 4-82A).

It has a lot of sound basic common sense about it but our experience with it is very limited so we cannot express an opinion but this is in no way a condemnation of this system at all; if you have to have reverse circulation then look at the eductor arrangement.

Mud Drilling

If you were to ask every person in the water-well drilling business "...What is mud?" almost all of them (at least ninety-nine percent) would say — bentonite, and yet that is a product which is banned from use in water wells in many high-tech countries because of its incompatability with water production (restricted yield because of formation blockage) and, surprise surprise, in the long run, its high cost per foot.

We will stop there and leave you to think about that until later when comparative costs of mud drilling will be discussed more fully. Here we will talk about mundane things like mud pumps and pits.

Fig 4-82A

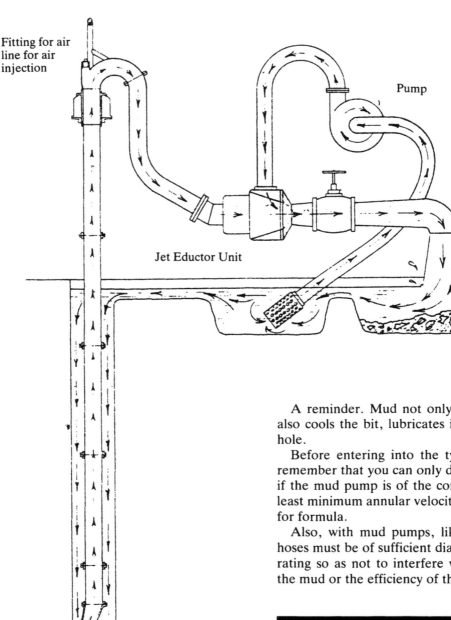

Fitting for air
line for air
injection

Pump

Jet Eductor Unit

The George E. Failing Company Reverse Circulation System
using patented Jet Eductor unit to create vacuum for circulation
– Fluid from the well and cuttings do not pass through
centrifugal pump, but go directly through Jet Eductor unit and
are discharged into pit.

A reminder. Mud not only cleans the hole, it
also cools the bit, lubricates it and stabilizes the
hole.

Before entering into the types of pump, just
remember that you can only drill holes efficiently
if the mud pump is of the correct size to give at
least minimum annular velocity — see Book Nine
for formula.

Also, with mud pumps, like compressors, all
hoses must be of sufficient diameter and pressure
rating so as not to interfere with the passage of
the mud or the efficiency of the pump.

Mud pumps

There are three types of mud pump in general use
today (plus maybe a few odd-balls about which
we will forget) and they are:-

1. *Centrifugal pumps* (figure 4-83A)
Here an impeller is spun inside a casing, the mud
being drawn in by creating a vacuum (or
priming), in a similar way to the reverse

Fig 4-83A

design have produced pumps of relatively high pressure. Our favourite centrifugal pump delivers 500 US gallons per minute against a 500 foot head.

Don't forget, it is volume that cleans the hole for you and the centrifugal pump is wonderful at doing such a thing at quite low cost. Why do you need high pressures anyway? Perhaps you don't, not that high anyway.

In our section on the "drill string" we went to some lengths to discuss tool joints on drill-pipe and how it is essential, when mud drilling, to have the biggest possible hole in them to overcome frictional losses imposed on the passing mud inside the pipe. You have to have enough pressure to overcome these losses and more.

Frictional losses in the annulus between drill-pipe and the wall of the hole can largely be ignored at average water well depths, but what about moving that great head of mud up the hole after you have changed a drill-pipe?

Let us first consider that we are drilling with water circulation — then the head of water required to be lifted is the difference between the top of the column of water and the top of the hole so, if your water is standing 2.31 feet below your conductor pipe then it will take one psi to move it (added to the frictional losses in the pipe) and if it is standing at 231 feet below surface (heaven forbid) then you will need 100 psi — why?

Because it resembles a "u" tube in that you have a column of mud left inside your drill-pipe when you shut of your mud pump to change the drill-pipe — the mud would have found its level outside and *inside* the drill-pipe (figure 4-85A).

Try this as an experiment. Get a length of hose about three feet long, bend it into a "u" shape and fill it with water. With your mouth blow down one side and you will easily push the water out of the other side. Now remove half the water and see how much more "blow" (pressure) you have to impose to move the water let alone push it out of the hose.

If you run a bit check valve (see drill string) behind the water (we recommend) you are even better off, because not only will it prevent the water finding its own level thus leaving the top of the column at its highest point, but it will prevent all those nasty bits of cuttings still in your mud (naughty) from washing into the bit and blocking

circulation pumps already described. The mud is then whirled round by the impeller and pushed out into the standpipe, thence down the drill-pipe to the bit. Then it goes up the annulus, through the channels (dropping cuttings as it goes) along the pits then, when clean (!!!) back into the pump — direct circulation (figure 4-84A).

Centrifugal pumps, sometimes known as rotary pumps, are designated by the diameters of the inlet, outlet and impeller. Therefore a 4x3x13 will have a four inch inlet, a three inch outlet and a thirteen inch diameter impeller.

These pumps are generally thought to be too low in pressure to be useful in drilling and yet, not only are vast numbers of rigs factory-mounted with them even at low pressure, but advances in

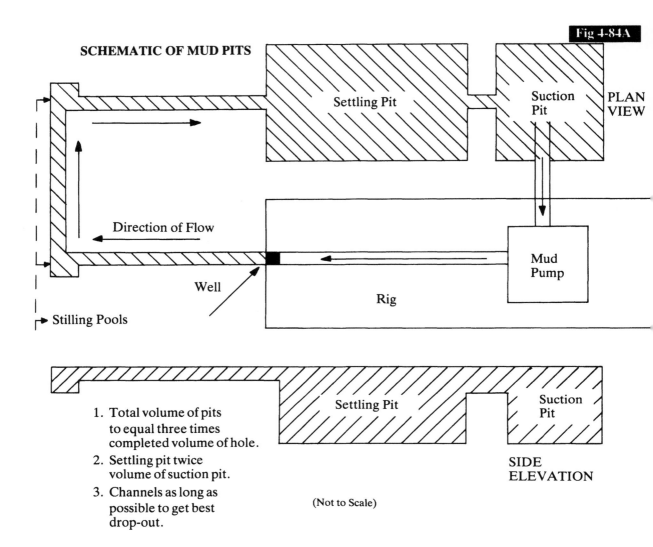

SCHEMATIC OF MUD PITS

Fig 4-84A

Settling Pit

Suction Pit

PLAN VIEW

Direction of Flow

Well

Rig

Mud Pump

Stilling Pools

1. Total volume of pits to equal three times completed volume of hole.
2. Settling pit twice volume of suction pit.
3. Channels as long as possible to get best drop-out.

Settling Pit

Suction Pit

SIDE ELEVATION

(Not to Scale)

it off. Here we are assuming no losses to the formations.

Your next question; what happens if you have lost circulation at depth, you have cemented off (no foam? — even naughtier) and now you have a deep column of water to lift? Simple — fill the annulus from the top of the hole and make up the difference.

Pressure requirements will increase with the increase in the specific gravity of mud (the thicker the mud the slower the drilling — remember?) and you can take it here to be in direct proportion, but remember — the higher the specific gravity (s.g.) the higher the pressure imposed on the walls of the hole. At depth this can be disastrous, causing formation fractures and

sometimes collapse — you did it, the mud didn't. It was you that mixed it or didn't recognise that you were making clay in the hole and take steps to thin it down.

All these things must be considered when calculating your pressure requirements.

So, a centrifugal pump can be a good idea in spite of its "low" pressure, but if your mud control is less than perfect it should never be used, because mud carrying even fine cuttings (especially fine cuttings) into the pump will wear the impeller very rapidly, seals on the shaft will deteriorate and eventually holes will appear in the casing. What is mud control? This:-

If you are going to use bentonite forget centrifugal pumps, as even a small lapse in

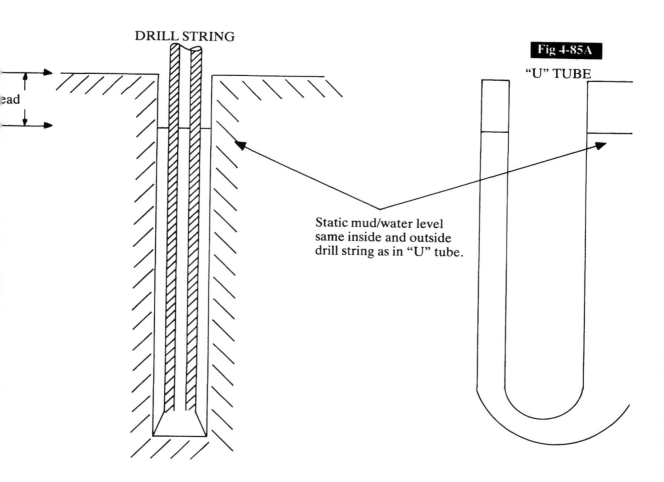

DRILL STRING

ead

Fig 4-85A

"U" TUBE

Static mud/water level
same inside and outside
drill string as in "U" tube.

vigilance over the mud will cause havoc in the pump, because bentonite is a dispersed system, and all those tiny bits of cuttings are separated in the mud and, as viscosity increases so it becomes more and more difficult to drop them out of the mud. Polymer mud however is a non-dispersed system and the cuttings tend to come together and are easier to drop. The mud pits, and the channels into them (figure 4-84A) should be of sufficient size to allow a long travel which gives a greater chance of "dropping out", even with polymers. As a rule of thumb (a guide) you should calculate the volume of your completed hole and multiply by three to give you the volume of mud needed in the pits. Then divide that by three, and make your suction pit to take one third and your settling pit to take the other two thirds. Channels should be as long as possible and the stilling pools (figure 4-84A) of sufficient size to effect good slowing of the mud travelling speed, thus getting maximum

time for drop-out. This is called retention time and a figure of 6–10 minutes is not amiss. To check the retention time put something that floats into the mud at the point of exit from the hole and time it to the point where it reaches the suction hose.

De-sanders or de-silters will help a lot as will a shale-shaker. Flocculants will also help. If you used all the items in this paragraph then you could run a centrifugal pump even with bentonite.

You will also need something to measure the viscosity of your mud and a mud balance to weigh it: a sand content kit is also useful. Neither of these items should be put away when you have made your initial mix because you should check your mud every hour at least. By relating one reading to another and taking your readings at more frequent intervals the mud will tell you what is going on down the hole.

If viscosity is down then there is a possibility of groundwater or even the presence of sand, which can tend to indicate a lower viscosity through a marsh funnel — more of that later. If, by taking a reading with the mud balance at the point of discharge from the hole and the point of entry into the foot valve (never put your suction hose on the bottom of the pit — it will pick up fines), you can tell whether or not you are dropping out cuttings because the mud weight will be lower at the point of entry, thus fewer cuttings — won't it?

ALL THESE POINTS ON MUD CONTROL APPLY TO ALL FORMS OF PUMPS.

And now a case history. This was a strange contract involving a British contractor and an Asian sub-contractor working in a West African country.

Our engineer was in the country doing something entirely different when a representative of the main contractor poured out his heart over the terrifying amounts of delay they were experiencing. We asked him what was causing the delays and he said that the sub-contractor had ordered the wrong sort of pump and was having to strip them (3 of them) almost daily, because of excessive wear on seals and sometimes even the impeller itself.

He was then asked our usual questions in such a case:-

Wear on the swivel packings? — Yes, a set of packings would last at best three shifts.
Mud used? — Why, bentonite of course.
Pit size? — Quite big really, about four feet by four feet and three feet deep.
Hole size and depth? — 9″ to an average of three hundred feet.
What is their mud control like? — He looked blank.

Our engineer was asked to meet the drilling superintendent on site and the conversation went like this:-

Us: There are alternative muds here that would help to cure your pump and swivel wear problems.
Him: I like to use bentonite because it controls the influx of groundwater until I am ready to take it.
Us (after explaining about minimum pit sizes): Your pits are very small.

Him: These fellows are too lazy to make them any bigger — it is too hot.
Us: Do you have instruments for measuring viscosity and weight of your mud?
Him: Yes, but they are back in town locked in the store.
Us: Do you use them?
Him: These fellows don't know how.

As we were talking the pit level was going down as the depth increased — which it does, of course, because the volume of the hole is increasing — and the crew were adding water and mixing new bentonite as drilling was going on.

Us: You really ought to let that bentonite hydrate before you use it, for at least twelve hours and, by the way, your mud pump is no longer pumping (also, his driller was being sprayed by mud from the swivel).

Drilling stopped for repairs to the pump and swivel and the representative of the main contractor, who was with us all the time, and our engineer, returned to town where we presented our report. The drilling superintendent returned to Asia very quickly and the last we heard he got a job as a drilling consultant with one of the major aid agencies.

Whilst most of the above is self-explanatory when taken in the context of the previous paragraphs, some points need clarifying.

By using bentonite to "control" the influx of groundwater he was in effect, forcing bentonite into the formations which, when it comes to cleaning out the mud during development of the well, remained in the smaller aquifers, so the yield from the well was very much reduced. If you ever see a water well that has been drilled with bentonite which suddenly dries up during development it is a safe bet that great chunks of wall cake have dropped off, blocking the screens altogether — a sure indication of bad drilling practices.

Because of the size of the hole being drilled the pits should have been nine times bigger than they were to be even mildly safe. The volume of his pit was less than the volume of normal run out channels and stilling pools — which he didn't have, by the way.

All muds have a period of hydration to allow

86

them to come to a *working* solution. With bentonite it is, in our business, a minimum of twelve hours and this can easily be seen if you check your mix for weight and viscosity first when it is mixed and then after twelve hours. Average hydration time for polymer muds is fifteen to twenty minutes.

Drilling is such a specialised business that people who interview other people for jobs do not normally understand the business and think that any old fool can do a dirty job like that. Anybody with a one percent knowledge of drilling can bamboozle his way through an interview conducted by people with no knowledge at all.

By the way, a centrifugal pump is sometimes used to boost a power pump by pumping mud from the pits into the inlet of the power pump.

2. *Positive displacement pumps* (figure 4-87A). Sometimes known as power pumps, piston pumps, reciprocating pumps, duplex pumps, triplex pumps, mud pumps (that's only half right). They can have one, two or three cylinders. They can be double acting or single acting and they can be driven by diesel engines, electric or hydraulic motors. Whichever way you look at it they pump mud (and many other things) and are the most widely used mud pumps of all.

They are designated by the bore of the cylinder and the stroke of the piston. Hence a 7½ × 10 has a seven and a half inch bore and a ten inch stroke. They are called positive displacement because that is their function. Mud is drawn (after priming) from the pits through the suction valves by the "draw" of the piston and a "positive"

Fig 4-87A

quantity of mud is deposited into a cylinder which is "displaced" out of the discharge valves into the standpipe, down the pipe etc. etc. to achieve direct circulation.

Pumps of a given stroke can have different piston diameters. The smaller the piston the higher the pressure and the lower the volume, the larger the piston diameter, the lower the pressure but the higher the volume.

Do you know what single acting and double acting means? Next time you see a mud pump of this type see if it has four valves per cylinder (two suction and two discharge) — if it has then it is double acting because it pumps mud on the forward and rearward strokes of the piston: two valves per cylinder, single acting.

So, a double-acting pump, which is usually "duplex" (two cylinders), will positively displace a cylinder full of mud on the forward stroke and a part full cylinder (the piston rod is there and will reduce the volume of the cylinder by the volume of the rod), on the rearward stroke. The pump is rated as a given volume at so many strokes per minute (bore × stroke × strokes per minute), but this is not strictly correct: it should be cycles per minute, there being two strokes (forward and back) in each cycle. Therefore when you count your pump strokes, count one cycle, which comprises one stroke forward and one back.

Single acting pumps? They are usually "triplex" (three cylinders) and they have (mostly) plungers rather than pistons and they pump into a common chamber, thus the peculiarity of having "odd" numbers of cylinders is smoothed out (in conjunction with a "damper"). They can reach quite high pressures.

Positive displacement pumps are more "long-suffering" than the centrifugal pump in that they will pump solids for a longer period before they run out of efficiency, but the result is far worse. Instead of an impeller and a few seals, you have pistons, liners, seals, valves, caps etc., etc., but your mud control *is just as important if not more* than with centrifugal pumps because *they are more expensive to run*.

It is idiotic to think that you can just keep passing contaminated mud through such a well engineered piece of equipment and get away with it. Half the time the inexperienced operator will not even know that the pump efficiency is down

and when it eventually becomes painfully obvious they will usually say "...change the pistons..." and forget that pistons are just links in a long chain of possible failures.

Mud control is absolutely essential and, as already said, is quite simple. You measure weight and viscosity at least every hour, having first decided on a good mud for the work in hand; mud control means a long pump life and has that same desired effect on your swivel and every other article of equipment through which the mud passes.

"GOOD MUD CONTROL MAKES LOTS OF HOLE".

As with all pumps there must be a safety (relief) valve in the system, usually in the discharge line immediately after the pump. If you are building up too much pressure in the hole (bit blockage?) then it is better for the relief valve to unload the system than to "burst" the pump and maybe cause injury. Make sure the setting is in accordance with the manufacturer's specification.

3. *Screw type pumps* (figure 4-89A).
These seem to be converted turbine submersible pumps and comprise a single screw turning inside a stator; the mud is fed in at one end (whilst the screw is turning of course) and is screwed out of the other. They are essentially low pressure (max. around 125 psi) but can deliver quite high volumes with a relatively low power input.

If your *mud control* is good then these pumps must be considered because they are reasonably priced and quite compact. Manufacturers seem to be able to fit them in the most unlikely places around a rig. But if your mud control is poor then forget this pump altogether because bits of sand etc., getting between the rotor (screw) and stator will spell total disaster.

This pump is measured by the diameter of the screw so you will want, for example, a three inch pump or two inch pump etc.

Pump maintenance

First and foremost, mud control will give a much longer life to your pump and an overall better cost per foot, because you will require less maintenance and breakdowns will be minimised.

Follow the manufacturer's maintenance

Fig 4-89A

schedules to the letter both for the pump and its prime mover. And don't forget the drive.

Running a pump dry leads to high wear — don't do it.

Never, repeat, never put your foot valve and strainer (suction hose) on the bottom of the suction pit — why? — if there are any cuttings left in the mud by the time it reaches the suction pit they will be fines, just the job for getting sucked up into the pump causing havoc. A pole across the pit to which you "rope" the suction hose is all that is needed.

Some good manufacturers will make the strainer in the shape of a large box the *top* of which has a fine mesh, all other sides being blank. You sit the box on the bottom of the pit and the suction is protected by the depth of the sides — we like that if it is well constructed.

Make sure your suction hose is well looked after (as with all hoses) especially when moving site — a pin hole will cause the pump to pump "air" which is not good for the pump — if it will pump at all. Also make sure the union with the pump is air tight.

And now for that hallowed subject — "drilling mud". This is where we all turn in to chemists (or alchemists) and wallow in ringing the changes in mud products to defeat that formation problem we are expecting.

We appreciate that in many countries, and we like to think that this book will be read in many, bentonite is the only viable source of drilling mud. In fact, we would go so far as to say this represents the majority.

OK. We can only say that we hope what has been said, and is to be said, is of some help. A case in point.

If you are working in a country where only

bentonite (or natural clays) are available it is reasonable to assume that the availability of pump spares is also a restricted product, therefore what we have said about mud control must be of considerable help.

"But we can't even get a marsh funnel or a mud balance" I hear you say, and what is our answer?

For a funnel and cup do the following. Get any sort of receptacle that comes to a neck, such as a plastic water bottle as you buy mineral water in or anything like it — it ought to hold about one litre. Cut the bottom (large) end off and put a bung in the small end (neck) and in the bung drill a hole about three sixteenths of an inch in diameter through it. Try to make it still hold the litre by thinly slicing the bottom off (figure 4-90A).

Now get a container of sorts, say an old food can that holds a bit less than a litre (or less than the thing you are using as a funnel). And now the procedure:-

With your finger over the hole fill the bottle with fresh water, then remove your finger and see how long it takes to fill the can and register the time. Now all you have to do is to compare that time with that of your drilling muds and you have your own source for checking viscosity. As long as you measure the same volume of mud as water then you will get a required comparison.

You really only have to use a Marsh Funnel because it is a standard and you can compare your results internationally — do you need that? — Of course not, not for this purpose anyway. What you need is to know what your mud is doing, and now you can.

For a mud balance all you need is something that will weigh a quantity of liquid fairly

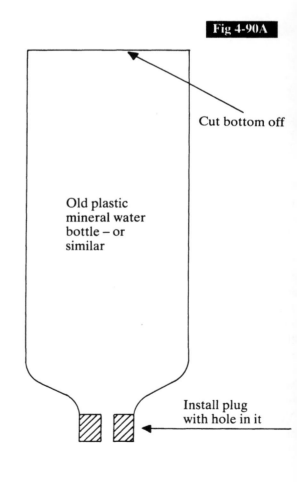

Fig 4-90A

Cut bottom off

Old plastic mineral water bottle – or similar

Install plug with hole in it

VISCOSITY SYSTEM

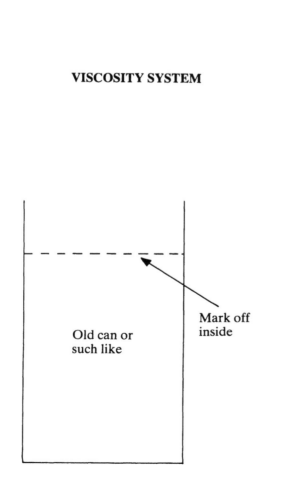

Mark off inside

Old can or such like

sensitively. So, weigh your litre or whatever of water and call that "1" (water has an s.g. of 1), then all you have to do is compare your mud weights and see how your mud is behaving provided you weigh the same quantity as with the water — QED.

Another thing to remember about bentonite is to use it sparingly if you have to at all. Overtreat (too much) and you create so many problems for yourself, your boss, and above all, the recipients of the water.

One final point about drilling with clay based muds (bentonite is a clay as is kaolinite etc.) they don't like saline water very much but there is a mud that isn't too fussy and that is attapulgite. If you have this problem (salty water) then that is a little tip.

And now a bit on polymer muds. The first thing to remember is that they are expensive to buy (but not to use — cost per foot remember?) and some can be even more difficult to mix than bentonite so, before you venture into this fascinating new world, make sure you have a good mud mixer on site.

Mud companies usually have good engineers available for consultation, but they mainly orientated towards the oil industry, so you should start off by saying that you are drilling for water because, another reason oil companies use such vast amounts of bentonite is to prevent the ingress of water into the hole, and the longer the better — we certainly don't want that, do we!

Then tell him (make sure he is an engineer not just a salesperson or if he is a salesperson that, he will pass on your request to an engineer). Your problems — say, shale stabilisation, or drop-out or whatever. You see, they all sell polymers but they are a small part of the business. However your friend will look through some catalogues, first blowing off the dust, and find you something. Watch out for toxicity, some oil field muds are hazardous, or indeed, poisonous.

The good mud companies even blend their own polymers and produce a ready made mud incorporating a variety of ingredients to suit a number of problems. For instance there is "Staflo", "Drispac" or "Custom Mud" (watch the first two: they are difficult to mix but good if well mixed). Perhaps you should mention these to your engineer and they will find an equivalent.

There are also separate polymers, such as CMC (carboxy methyl cellulose) a great viscosifier and hole stabiliser and much used to control filtration and cuttings removal. HEC (hydroxy ethyl cellulose) much as CMC but excellent in saline conditions — even to brine. There is also XC (xanthan gum) a very superior viscosifier (but expensive) and suspension medium.

We've said it before and, I'm afraid I have to say it again, the best action is to take advice from a good engineer and find out what is available in your country because:-

1) 4–5 pounds of polymer (general figure) per 100 gals. of water has to be compared with up to 50 pounds per 100 gallons of bentonite for a decent mud mix — think of the difference in transport costs alone.

2) You need 15–20 minutes of hydration time for the polymers to yield into a good mud from mixing, as against a minimum of twelve hours for bentonite; think what that saves you in time alone without even considering dumping bentonite mud when it becomes contaminated — that's another twelve hours at least.

3) You will still have to add a polymer to your bentonite to overcome problems anyway or, for some conditions, buy special clay products.

Of course, all this is academic if you are drilling in the ninety percent of conditions where air or air/foam/polymer will work better.

Don't forget, you only *have* to use mud in preference to air/foam/polymer if you have to build up hydrostatic head — for example in water bearing running sands.

Polymers are also available in liquid form, but make sure they don't contain any contaminants that might cause problems in a water well.

Enough of all that. What about weight materials for use where you have to build up high specific gravities to overcome exceptional formation pressures? There are a hundred trade names for this but most of them come down to barites (barium sulphate) which it is quite common to use with bentonite. With a natural specific gravity of upwards of 4.2 you don't have to use much to kill artesian pressure — work it out, but take advice.

When drilling with polymer muds, salt is a good weighting material and three pounds to a gallon (imperial) of mud will give you 10 ppg and more.

In known areas of high formation pressures weighting material is the only mud product that should be stacked near the rig and the last thing you load for transport to the next site whilst finishing a hole. You never know when you will need it.

Let's go on a bit to flocculants. Know what they do? They will help you get rid of cuttings from your mud. Remember what we said earlier about treating mud chemically for this purpose? Well, these are what you use if you can't "drop-out" any other way — talk to your mud engineer.

How about water loss, or lost circulation? (Remember the difference?). There are any number of products for this. Most will be either fibres, shells from nuts or flakes of something (flakes of mica, for instance). But let us tell you a little story that should be helpful to all you drillers working in most of the countries of the world. Check with your client before using — but it shouldn't be a problem.

We were working in East Africa, way out in the bush, drilling volcanics (see basic geology) and my client had not supplied any LCM (lost circulation material). It was the old old story: if you haven't got it you will need it, and we got it in a big way — lost circulation that is.

We sent a truck out to collect camel, donkey and horse manure and we set about crushing it into its natural fibres (the crew wouldn't touch it — the first time anyway). We mixed it with the polymer mud, with which we were using a special lubricant, and, hey presto, we drilled on nicely. If you have what you consider to be wide fissures in the hole, then drop in manure from goats or sheep uncrushed.

That is really not a joke: we have used it for donkey's years — that last remark was meant to be a joke. Whilst this might seem distasteful to your client it is cheap, isn't it, and mostly available, isn't it, and I don't think the chemist chappie will mind too much — ask them.

It beats the hell out of cementing lost circulation zones.

LCM that returns from the hole will be taken out by the shale shaker, so clean it off and put it back in the pits if you are still suffering.

Also watch your strainer. That can get blocked by the LCM.

Note: We have talked about wear on mud pumps but the same thing applies to any pump. If you put something through it that is alien to its designed function then in will react by breaking down, won't it. Let us take an example of the humble hand pump which is installed world-wide by the million.

The siting, drilling, screening, gravel-packing and development of a water well plus the installation of the hand pump (also applies to submersibles) is an entity and should be treated as such. Any deviation from the norms will affect the pump by, say, solids passing through it causing wear, hole deviation causing heavy pump action at surface (or damage to the rising main and/or the pump rods), incorrect understanding of the water levels (static, dynamic and pump setting) affecting good pumping throughout the year etc. Well development is the main problem here, remember, you can't develop a well and settle a gravel pack by just pumping large quantities of compressed air into the well — it doesn't work. See Book 6.

5 Casing and Cementing

Foreword to Book Five

Casing and cementing have been put together because they go together.

Did you know that in the "oilfield" they sometimes call in special crews to run casing and will almost always call in specialist companies to cement? The latter we can understand because it requires a lot of specialist equipment and personnel, but to run casing — that is something else.

Anyway, we poorer souls have got to do it ourselves so let us hope that what follows will help you in your endeavours and make it that much easier.

Now read on.

Fig 5-94A

Before entering into any discussion about casing, steel or the preferred plastic, we have to mention again that thorny subject of thread protectors and careful handling.

It is very rare indeed to walk onto any well-site and find threads protected on casing (or drill-pipe for that matter) and it is most common to see great dents in the threads and other untold damage. Figure 5-94A shows how damage can occur even with protectors if they are badly designed.

It is also common, when casing is about to be run, to find crews with hammers knocking out dents from threads and scraping layers of rust off them and making a half-hearted attempt at greasing and trying to make couplings fit.

We have seen a truck load of casing arrive at a well-site, the crew pulling the casing half off so that one end is on the ground and the other still on the truck, and the truck drive away leaving the casing to crash on the ground and *together*.

The time saved by looking after casing and, for that matter, any thread, pays dividends in the end.

There are hundreds of different types, sizes, thicknesses, diameters and grades of casing and any combination of any of them and a sackful more, so what we have tried to do is boil all this down into those most commonly used in water wells, and figure 5-95A lists the results.

Let us start off talking about plastic because this type of casing has become a boon to the industry, not only because it is so easy to handle, but also because it appears to have an indefinite life in the ground even under the most trying conditions.

We first put dear old PVC casing down in the mid sixties where the ground water would have eaten steel away in a year or two — even stainless steel with its inherent impurities — and it is still there.

The trouble with PVC (poly vinyl chloride) was (and, unfortunately, still is) its nasty habit of splitting, especially in the box end thread, and those splits extend very rapidly — too brittle. Then came the wonderful ABS (acrylinitrite butediene styrene) that you could drive a truck over and would bend, and return to normal, and

upon which we have seen a lightish rig standing. But that was too expensive.

About the same time came GRP (glass reinforced plastic, or fibre glass) which was also expensive, but the biggest drawback of which was that dust raised from it during machining for threading was a health hazard.

The threading of casing is much preferred as "bonding", grommets or grub screws can prove somewhat unreliable at depth. Don't forget, sometimes the casing has got to come out again and casing with poor connections can come apart with a little pull from the rig.

Then along came polypropelene (ppl) which seems, at the moment, to be the most suitable answer to the above problems and many casing manufacturers use this as the basis for their products.

Plastics — say in class D or even C — can be actually threaded into the wall of the casing leaving it OD flush, but the strongest method is to externally upset the raw material and use the upset end as the box end thread. Whilst some manufacturers favour a fine thread similar to, say the API, STC (short thread coupled) of 8 TPI others (we also) prefer the buttress type — (figure 5-95B).

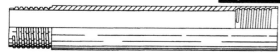

Fig 5-95A

Fig 5–95B

Fig. 2—Flush Butt Joint Casing

TABLE B — B.S. SPECIFICATION 879 WATER WELL CASING
SCREWED FLUSH BUTT JOINTS — Fig 2

Nominal bore in	Tube					Thread (i)		Max Bit to pass thro' (ii)
	Outside diameter in mm	Wall thickness in mm	Actual bore in mm	Weight Steel Tube lb/ft kg/m	Weight ABS Tube (iii) lb/ft kg/m		Thd/in Pitch mm	in mm
4	4½ 114.3	0.312 8.0	3⅞ 98.4	14.0 20.9	1.89 2.81	4	6.35	3¾ 95.25
5	5½ 139.7	0.312 8.0	4⅞ 123.8	17.4 25.9	2.55 3.80	4	6.35	4¼ 107.9
6	6⅝ 168.3	0.375 9.5	5⅞ 149.2	25.0 37.2	3.60 5.36	4	6.35	5⅝ 142.9
8	8⅝ 219.1	0.375 9.5	7⅞ 200.0	33.0 49.1	4.76 7.08	4	6.35	7⅝ 193.7
10	10¾ 273.0	0.438 11.0	9⅞ 250.8	48.0 71.4	7.41 11.03	4	6.35	9½ 241.3
12	12¾ 323.9	0.438 11.0	11⅞ 301.6	57.3 85.3	8.85 13.18	4	6.35	10⅝ 269.9
13	14 355.6	0.438 11.0	13⅛ 333.4	63.1 93.9		4	6.35	12¼ 311.2
15	16 406.4	0.500 12.7	15 381.0	82.0 122.0		4	6.35	14¾ 374.7
18	19 482.6	0.500 12.7	18 457.2	98.8 147.0		4	6.35	17½ 444.5

Note: (i) Square form, parallel thread.
(ii) Maximum bit diameter is the largest standard bit which will pass through a diameter ⅛". 3.175 mm less than the actual bore of the casing.
(iii) ABS – Acrylonitrile Butadiene and Styrene – an extracted plastic tube having approximately the same wall thickness, screwed to B.S.879 and about one eighth the weight of steel tube. A tough, impact resistant, temperature stable, non-toxic and non-taint material.

TABLE C — B.S. SPECIFICATION 879 WATER WELL CASING
SCREWED & SOCKETED JOINTS — Fig 1

Nominal bore in	Tube				Socket		Thread (i)		Max Bit to pass thro' (ii) in mm
	Outside diameter in mm	Wall thickness in mm	Actual bore in mm	Weight Plain Tube lb/ft kg/m	Outside diameter in mm	Overall length in mm		Thd/in Pitch mm	
4	4½ 114.3	0.25 6.4	4 101.6	11.3 16.8	5⅛ 130.0	4½ 114.0	10	2.54	3⅞ 98.4
6	6⅝ 168.3	0.312 8.0	6 152.4	21.2 31.5	7¼ 184.0	5 127.0	10	2.54	5⅞ 149.2
8	8⅝ 219.1	0.312 8.0	8 203.2	27.9 41.5	9 5/16 237.0	6 152.0	8	3.175	7⅞ 200.0
10	10¾ 273.0	0.375 9.5	10 254.0	41.6 61.9	11 7/16 291.0	7 178.0	8	3.175	9⅞ 250.8
12	12¾ 323.9	0.375 9.5	12 304.8	49.6 73.8	13⅝ 346.0	7 178.0	8	3.175	10⅝ 269.9
13	14 355.6	0.375 9.5	13¼ 336.6	54.6 81.3	14⅞ 378.0	8 203.0	8	3.175	12¼ 311.2
15	16 406.4	0.375 9.5	15¼ 357.4	62.6 93.2	16⅞ 429.0	8 203.0	8	3.175	14¾ 374.7
18	19 482.6	0.375 9.5	18¼ 463.6	74.6 110.0	20 508.0	8 203.0	8	3.175	17½ 444.5

Note: (i) 55° V thread form – taper 1/64" per inch on diameter.
(ii) Maximum bit diameter is the largest standard size bit which will pass through a diameter ⅛". 3.175mm less than the actual bore of the casing.

Inside the bottom of the somewhat flimsy casing we welded some bits of old (thin — that's all we had remember) sheet steel in the shape of a taper, the ID of which was just bigger than the OD of the tool joints: we ran this into the hole attached to the 6″ casing.

Let us clear up another point before proceeding and that is the international name for casing which has a separate double box threaded adaptor for connecting two pin threaded joints of casing. It is called a coupling, not a socket and certainly not a collar. Another sermon over, so let us carry on.

It is unlikely that plastic casing will have couplings. It will either be externally upset or OD flush.

Did you know that, internationally, plastic casing comes in 5.8 metre lengths? Why? — Because it will fit inside a 20′ shipping container. Remember that, the next time you have a critical length of casing to run.

The storage of plastic casing is important because, badly done, it will "bow" and "twist" and will therefore be difficult to run down a hole in spite of its flexibility. Your supplier will give you more definite information but we like to see the bottom layer supported along its length by a number of wooden slats to prevent "bowing", the next layer lain crossways across the first (figure 5-96A) the next crossways across the second and so on, but never more than four layers high.

Even supporting under the upset end will cause distortion, so don't put a slat under there.

Keep it in shade as much as possible because there is a certain amount of softening at high

Fig 5-96A

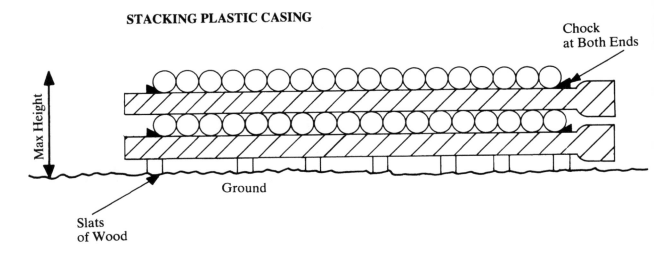

STACKING PLASTIC CASING

Chock at Both Ends

Max Height

Ground

Slats of Wood

temperatures which could also lead to distortion.

Here is a little tip about screwing plastic casing together. As you will know it can be very difficult when the threads are dry but, if you have foam on site, then dip the pin thread in it and you will find it will screw together beautifully. No foam? The village shop will have some shampoo or the like.

The "hoop" strength (lateral) of plastics is quite high, therefore slips can be used in the working/ rotary table but the use of chain tongs or stilson (pipe) wrenches is not recommended because they will impose undue loads on the casing at isolated points; a "wrap-around" (multi-point) tong is by far the best; specialist tongs are available.

Steel casing should not be confused with any old steel tube that you can buy at any old stockist, there are stringent standards laid down by such august bodies as API, British Standards Institute, etc. and these standards govern steel spec., roundness, straightness, threads, couplings etc., and, here we go again, thread protectors.

All these points are of the utmost importance if you are going to drill to standards. Figure 5-97A gives you some idea of casing standards and figure 5-95A which bit sizes will allow the passage of which casing and which casing will pass which bit sizes — all so important.

In a recent job we had to do in a country where the only available "casing" was GI (galvanised iron) pipe, the threads were so poor we had to

tack weld each joint as it went into the hole and the straightness so poor that it made contact with the hole all the way down.

Another little tip here. If your casing suddenly stops as you are going down when there is plenty of hole below it, it probably means you have a straightness problem either with the hole or with the casing. Ease it back slightly, give it a turn with your chain tongs (allowed with steel casing) and almost always it will go.

Making up good quality casing that has been looked after is so easy. The time saved in running it will more than save the difference in cost between that and rubbish.

Again, our preferred thread is buttress. It makes up a dream but is expensive and really far too good for all but the deepest holes (but not for plastics — remember!). Any "approved" thread properly executed will give you good service.

Steel specification is according to depth to be cased. The greater the depth the higher the specification, as also applies to threads. Your supplier should be able to give this information to you but, as a guide, figure 5-97A gives you some idea of strengths etc.

A little story here. An engineer was walking on to a well-site one morning and the driller was happily waving "good morning" — at that moment the engineer was horror-struck as the casing was slipping ▪down below surface — the driller was unaware of what was happening.

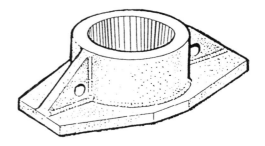
Fig 5-97B

CASING & DRILL PIPE STEEL SPECIFICATIONS

Fig 5-97A

Drill Pipe Casing	Pounds per Square Inch		
	Min Yield	Max Yield	Tensile Strength
H40	40,000	80,000	60,000
J55	55,000	80,000	75,000
K55	55,000	80,000	95,000
N80	80,000	110,000	100,000
Grade —	75,000	105,000	100,000

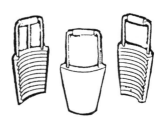

Apparently, he had almost finished running the casing in but a little sediment was holding it up. He had his drill-pipe inside and was washing the casing out — a perfectly normal procedure — except that he had removed his clamps. Thus the casing was free to run. This was an idiotic thing to do. Fortunately it did not go too far into the hole and fishing was easy — but it might not have been.

Casing handling tools

Your first choice of tools for holding the casing in the working/rotary table is a good set of slips with good sharp teeth. Slips are positive and the more weight the tighter they get.

Clamps for this purpose are time consuming, thus very expensive to use. Clamps can take as much as fifteen minutes to screw and unscrew for each joint of casing so, for an average water well rig twenty joints of casing at fifteen minutes per joint translated into cash will probably buy you a set of slips.

If you don't have a slips bowl in your table then another forty joints run will probably buy you a sub-table spider — see figure 5-97B for the picture.

Wrap-around tongs (figure 5-97C) are always best to use for tightening (and loosening) casing,

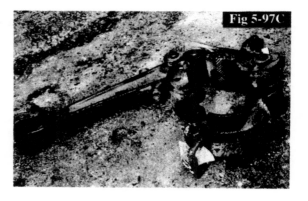
Fig 5-97C

especially plastics, although a good set of chain tongs have been used for a century or more and cannot be ignored when using steel casing.

For lifting the casing off the ground (and putting it down again) you have two main choices:-

97

Elevator

See figure 5-98A. These can only be used with externally coupled casing because they hold the casing from under the coupling. They are excellent tools and need not be heavy in these modern times. They are not, however, very good with plastics because the plastic can float in a wet hole and can go upwards through ill fitting elevators, so what do you use then?

Lifting plug (bail)

Figure 5-98B is a typical casing lifting plug and you will remember we discussed such a tool, under drill string, that was for lifting pipe and collars.

The construction details are the same, a suitable thread for the casing, a suitable bail for your winch hook, both connected by a shaft that works onto a good thrust bearing — all of sufficient capacity for the casing weight in hand. All lifting plugs should have a relief hole drilled through them to equalise pressure inside and outside the tools — you could "swab" the hole even when casing in loose formations.

If your lifting plug for the drill-pipe is of sufficient capacity then a series of adaptors from that to your casings is all that is required.

Ideally you should have two lifting plugs, one in use and the other in a waiting joint of casing — it saves an awful lot of time.

The lifting plug is good for plastics and essential if you are running OD flush jointed casings, plastic or steel. What is flush jointed? It means that the thread is cut directly into the wall of the casing (no upset) therefore the outside is the same diameter all through.

There is also flush coupled, where casing is joined by a coupling that is flush on the OD but slightly smaller on the ID All reference to flush jointed in respect to handling applies also to flush coupled.

Flush jointed casing is mainly used in diamond drilling but can be useful in a tight situation.

Your casing must always go into the hole with threads fully tightened.

Another little tip when running plastic final strings. As you know, you should always have at least one joint of plain casing at the bottom of the string which should be plugged to prevent ingress of solids — it is known as a sump. It is there to catch any settlement from the groundwater.

Such a string will float in the water in the hole and, in fact, is likely to bob up and down — fill it with water and you will be able to handle it better until the screens allow water to enter the string.

Here are a few other bits and pieces that you should have with your kit of casing tools:-

SIMPLE CASING LIFTING PLUG

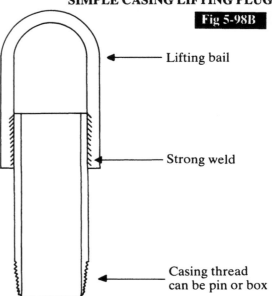

Fig 5-98B

- Lifting bail
- Strong weld
- Casing thread can be pin or box

Fig 5-98A

Casing shoes

In normal, run-of-the-mill casing strings it is hardly necessary to put a "shoe" on the first joint. Maybe just a coupling, fully tightened. But if you are expecting to have problems requiring, say, a bit of driving or turning then use a simple plain, bevelled shoe for driving, and for turning, a serrated shoe is OK.

If you are running an OD flush string which is going to be drilled in then firstly you will need a casing spinner, (top drive rigs only) which is a sub. between the rotary head and the casing, also a substantial shoe, normally serrated with either hard facing or tungsten carbide inset: sometimes even diamonds.

Figure 5-99A illustrates a range of shoes and most are available off the shelf from suppliers.

Pup joints

What are they? They are short lengths of something, in this case casing. Very handy, and you ought to have, say, 2', 5' and 10' pup joints. Not only will they sometimes make up a string on odd measurements but they will give you that little bit of flexibility if you find yourself in a spot of bother with obstructions. You've got pup joints of drill-pipe haven't you? No? Well get some — they are just as handy.

Landing joint

We like to have one of these made up and ready at all times — it is for cementing and is a joint that enables you to locate the top thread of your casing string accurately for when you are planning a cellar.

This is the one occasion when you require casing clamps because, as already mentioned, if you are going to cement the casing in you will not have any upward movement when dry to release slips. If you have upward movement, you are in big trouble.

So, you have your landing joint complete with clamps and you will need a simple cementing head — see later. Don't forget, casing is seldom exactly the same length, so you must always measure each length you run, and mark it — this also applies to your landing joint.

Casing centralisers

They should *always* be used. Their presence when cementing casing, gravel packing screens and the like is essential *if a good job is to be done*; bridging during these latter operations must be *avoided* — use them.

SITE INVESTIGATION EQUIPMENT

CASING

DESCRIPTION	ILLUSTRATION	STANDARD DIA.
Casing British Standard Water Well Casing Flush Butt Joint to BSS 879 Table 3 Hot Finished Seamless Steel Tube 35 T.T. (12" dia 28-35 T.T.)	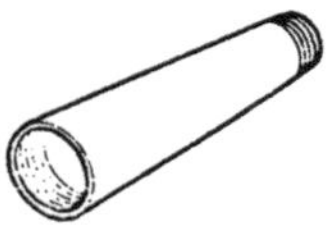	6⅝ O.D. × 5⅞ ID 168mm × 149mm 8⅝ O.D. × 7⅞ ID 219mm × 200mm 10¾ O.D. × 9⅞ ID 273mm × 250mm 12¾ O.D. × 11⅞ ID 324mm × 302mm
Casing Thread Protector Female (For Male End)		
Casing Thread Protector Male (For Female End)		
Drive Head Full Bore with Clevis and Pin		Also available in Box Thread
Drive Head Slip in Type		Rarely used
Casing Shoe (Plain)	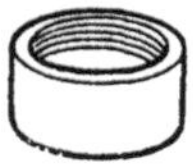	Also available in Pin Thread

SIMPLE HANGER Fig 5-100A

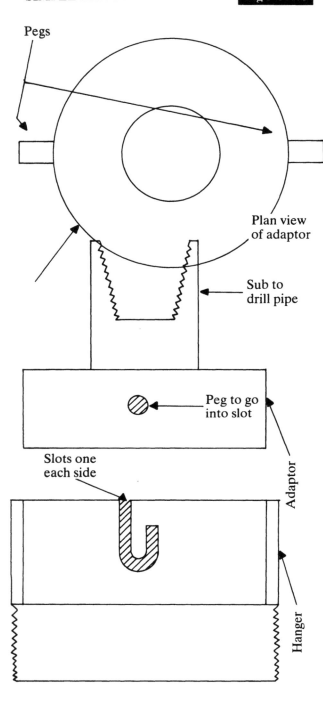

Pegs

Plan view of adaptor

Sub to drill pipe

Peg to go into slot

Slots one each side

Adaptor

Hanger

Casing hangers

These are used for placing casing strings inside a hole — that is, not terminating at ground level. They are almost always run on drill-pipe.

They can be simple (you can make them on site) or complicated, costing mega-bucks. Really, unless you are using great casing weights, the simpler the better.

Figure 5-100A plus 5-101A shows two types in general use. The first is a left hand thread adaptor to your drill-pipe so that, when you have the casing string settled you can forward rotate your drill-pipe, thus undoing the hanger and not the drill-pipe or casing.

The second is the simplest and cheapest (no expensive threading) and is merely a piece of casing with slots cut in bayonet fashion so that an adaptor (recoverable) on the drill-pipe, with two (or more) prongs on it, fits into the bayonet for lowering and, when in position, a slight lowering and turning of the pipe will release the adaptor. Isn't that clever?

Casing welding jig

This is for use when you are welding joints together and helps you to get the casings welded together straightly. Figure 5-101B shows just such a jig, which looks like a deep casing clamp with four windows cut in it. The weld prepared casings are offered to each other and the jig is tightened around them with the joint to be welded showing in the windows; this tends to pull the two joints of casing together. Tack weld through the windows, remove the jig, and complete the welding.

Screens (filters)

This is where the arguments start, because the use of screen in the hole to filter the water is a subject of such controversy that the only consistent arguments we have heard have, in fact, been wrong — even from manufacturers.

Screens can be plastic, steel, stainless steel, can be flame cut (ghastly) saw cut, drilled, wedge wire wound, louvred, *punched, drilled and mesh*

THREADED CASING HANGER Fig5-101A

Adaptor to drill pipe

Left hand threads

Casing thread

Fig 5-101B

wrapped, have filter material bonded on — and still we haven't finished.

They can have weld-prepared couplings or screwed, or dowelled, or just about any other mechanical contrivance known.

In actual fact, only one is a filter and that is the stainless steel wire wound. We cannot include the bonded type because that is two things in one. So why is the wire wound the only true filter? (And this is where we start the arguments).

Because, to use them correctly, an analysis of the fine sands in the aquifer should be taken and then the slot size chosen to suit. If you are going to use a gravel pack you should not be using wire wound screens but one of the less expensive alternatives. Yet people will continue to use them with gravel pack forever — and waste vast amounts of hard earned cash.

We still prefer to use plastics with a sawn slot (thousands of them) about 0.020″–0.040″ wide and a bunch of correctly graded gravel, and talking of gravel, it must be gravel that is rounded and not stone chippings. The size of gravel is commensurate with the slot in the screen

combined with the sieve size of fines in the hole. This will usually be specified by your client.

Another point here. To run a gravel pack the hole should be at least 5″ bigger in diameter than the outside diameter of the casing couplings, giving a minimum 2½″ annulus; anything smaller and the restricted annulus will cause bridging of the gravel.

Screens should be set in the proximity of the producing aquifer, or aquifers if you are screening at different levels, and if you can't be exact err downwards because the dynamic water level will be above the aquifer won't it? Don't confuse rest water level, where the water "rests" in the hole without the pump in operation, and the dynamic level which occurs after drawdown of the water in the hole.

Simultaneous drilling and casing

Before starting on this one, let us consider one very big point, and in fact make a statement of fact which we have proved over the years to our satisfaction and to that of all but the most obstinate of our clients who think they know everything and yet cannot interpret what is obvious.

If you have the equipment (simple) and materials (normal) at your disposal, then it is easier and vastly cheaper to drill a hole and put casing in than to go for one of these simultaneous systems, bearing in mind we are drilling water wells.

In civil engineering works, minerals exploration and even on extremely rare occasions in our industry, it is necessary to drill and case simultaneously, but for us, it is so rare, that if you follow the rules we have set down in these books, you won't come across such an occasion in a hundred lifetimes.

We have had only two examples of using these systems in water wells. Firstly, drilling near horizontal holes (yes, for water) in sediments (starting the hole) and the other when we were asked to drill across an underground river.

A story. We did a very successful series of tests in boulders — considerable amounts of boulders — and our client, an immensely experienced (?) man said we hadn't been drilling boulders at all and yet the evidence was there for him to see — he reverted to simultaneous drilling and casing, increasing time taken to drill the well by up to five times and his costs by much greater than five — and that was in a drought area.

There will always be an exception to the rules of course, and here it is — the old cable percussion rig, driving its casing as it goes — an admirable though somewhat dated system.

Anyway, the preamble is over. Drilling and casing at the same time seems to fall into two categories:-

The first is where the casing and rotary drilling tools are coupled to the rotary head (not for rotary tables unfortunately) and they are both rotated; the casing having a drilling shoe fitted. This is mostly used in those formations with a hardness suited to rotary drilling. Flushing goes down the inside of the drill-pipe and up the annulus between the drill-pipe and the casing; a certain amount of the flushing is allowed to pass outside the casing for cooling and lubrication of the casing shoe. A special adaptor is needed to pass the cuttings out of the system at the rotary head (figure 5-103A)

The idea here is to drill through the overburden with the casing and tools together and then, when "safe" ground is reached, go forward with the drilling tools leaving the casing behind. The actual drilling tool can be a drag bit, rock bit or hammer.

Think about our problem with the underground river for a moment. We drill beyond the river using simultaneous drilling and casing, do our analysis of the cuttings, put the required grade of stainless steel wire wound screen, on a packer, in the hole inside the casing and pull the casing back, exposing the screen to the water, just far enough to support overburden.

The other drilling/casing method is where you use an eccentric bit on the tools, usually a hammer. The casing does not drill but is eased down behind the bit; reverse rotation closes the eccentric bit for withdrawal through the casing. Figure 5-104A gives a schematic view of the system.

There is another system, but very rare, where a disposable bit is used with a bayonet fixing to the drill-pipe, which is bigger than the casing. When the hole is finished a quick part turn of the tools releases the bit, and the tools withdraw leaving the bit in the hole.

There is a place for all these things — that is why they are included, but your first choice must be to drill a hole and put your casing down, in that order.

Site layout

Keep your casing neatly stacked, the right way round, and within easy reach of the rig — you don't want to have to move it around unnecessarily — it is heavy.

Put it on wooden slats, keeping the layers of casing separate so that you can apply elevators or a lifting plug easily.

Threads should be greased and prepared before running, it is wrong to have someone cleaning off rust, banging out dents and greasing as each joint is offered to the rig — it happens.

When a joint is being pulled up on the winch

don't let the "loose" end drag along the ground — you will only have to clean it again and probably knock out another dent; this applies even if you have thread protectors. Support it somehow, even if it is only a turn or two of rope around the loose end and people pulling against the travel of the winch, thus elevating it.

DUPLEX HEADS

Heads incorporating threads in the heavy duty system are manufactured for connecting to various top drive rings.

We have various flange arrangements used on Hands—England, Duke & Ockenden, Hydraulic Drilling Ltd., etc.

When ordering it is only necessary to state: Rig/casing size/rod size, whether rock bit is external or internal cutting.

Before talking about cementing, let us discuss something that is an essential "first step" preparatory to a cementing operation, or indeed, to any casing operation: it might seem time wasting but it is, in effect, a time (and money) saver. It is called a:-

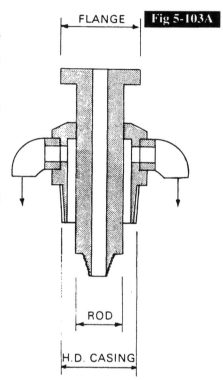

Fig 5-103A

CUTTING SHOES

SIZE OF CASING	A
$3\frac{1}{2}''$	$3\frac{11}{16}''$
$4\frac{1}{2}''$	$4\frac{11}{16}''$
$5\frac{1}{2}''$	$5\frac{11}{16}''$
$6\frac{5}{8}''$	$6\frac{3}{4}''$

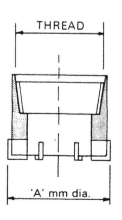

All tungsten inserts are flush on inside diameter of shoe.

103

Fig 5-104A

SIM-CAS 4
Simultaneous Casing System

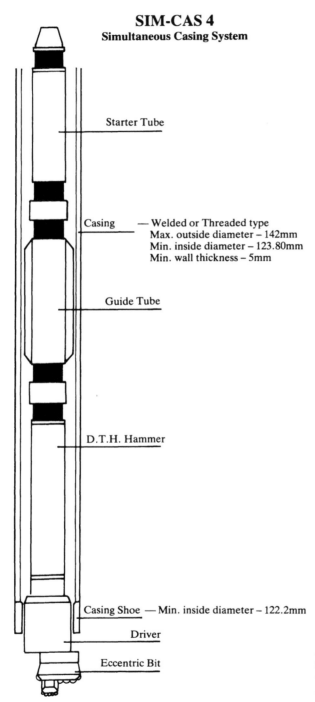

Starter Tube

Casing — Welded or Threaded type
Max. outside diameter – 142mm
Min. inside diameter – 123.80mm
Min. wall thickness – 5mm

Guide Tube

D.T.H. Hammer

Casing Shoe — Min. inside diameter – 122.2mm

Driver

Eccentric Bit

Wiper trip

"Wiper" means clean and "trip" means running in and out of the hole — doesn't it! So it is a method of knowing whether or not the hole is clean.

When you have reached the required drilling depth, wash out the hole with the bit near bottom until your mud comes clean.

Trip out of hole to the bit and wait for about thirty minutes, during which time you clean the bit and get your mud into good condition (We hope you're checking it at least once each hour).

Run your tools back into the hole and feel for any settlement. If there is, increase the viscosity of your mud and clean for at least thirty minutes. You might have to make a second wiper trip if hole conditions are bad, or even a third, to be sure that all is well for casing and cementing.

If you aren't using mud the wiper trip is still necessary and you can increase viscosity in the foam/polymer rotary system by adding more polymer and, with pure air, adding foam/polymer.

Cement

Don't forget — as soon as cement is mixed it starts to harden and you have to make all preparations for getting the cement slurry in place and your pump, line and tools free of it before you get into trouble. Have you ever seen a mud pump full of hard cement? No? Well we hope you never will because it takes an awful lot of chipping out.

We are presuming that we are talking about cementing in casing by filling the annulus between it and the walls of the hole or the annulus between two strings of casing. Just filling a hole with cement requires all the normal preparations but can be limited to, say, pouring, or the "tremie" methods that are to follow.

We again presume you have access to normal, good quality "Portland" cement or other such special cements as developed by oil companies, for instance, for their own special needs — always take advice if you are in any doubt at all.

There are also many additives available for your special needs, such as accelerators, decelerators, expanders, water softeners (ph adjusters) etc., etc. — take advice. In areas of high salinity don't forget Sulphate Resisting Cement (SRC) and there are other grades of cement for other purposes which will make your well that much better, remember — TAKE ADVICE.

One additive we would like to talk about is our old friend bentonite — why do we use it? Because, properly mixed,(and you always mix the water and bentonite before adding the cement), it allows better suspension of the cement particles and, more importantly, it makes the cement "run" better into cracks or down tight annuli.

If you need this assistance then add to the water 2–3lbs of bentonite for each sack of cement (94lb sack) you plan to use. Polymers may also be used for this purpose but in a very much reduced proportion, say, ½lb per sack or less. It is trial and error and you will learn from your first major cementing job.

Sand, as an additive, should be used with caution as it will have exactly the opposite effect to bentonite — it will encourage bridging, although it does make up bulk. Sometimes you need to make cement bridge, especially where large cavities are concerned or in open holes, so take advice, because a compromise might be required: you might even have to use concrete in very difficult circumstances.

Let us make a rule here, don't use sand in your cement if you are to cement casing annuli.

For other works it might be OK and you should add sand in varying proportions according to the job in hand; the grain size of sand is also important, the bigger the grain (coarse sand) the greater the bridging effect, and so on.

Mixing cement

Always mix enough cement for the job that is to be done, even if it requires an additional pit in which to do it — in other words cement all in one go otherwise you could get a "fissure".

Use a "proper" method of mixing — something like a good jet mixer as mentioned under "Auxiliaries" — a couple of people with shovels cannot get the lumps out.

Get the proportion of water (bentonite) and cement right: too much water and you will get separation and shrinkage, too little and it won't pump or pour.

Have a good quantity of fresh water available (before you start) to wash out your pump and tools.

Have all pump dismantling tools available (before you start) because, like it or not, you are going to have to take your pump apart and clean it *immediately* you have finished cementing.

Have all your cement assembled at the mixer (or whatever) to complete the work. For this you must calculate the volume of the annulus to be cemented and, to allow for fissures, washouts etc. add some more. We like to see about 30% extra; this should leave a nice little plug in the bottom of the casing to be drilled out. If you know of voids, make allowances.

We presume you have your casing landed — allowing the shoe to be 2–3' off bottom — your landing joint clamped, and your cementing lines out with the cementing cap (see later) standing by in case that is to be the chosen method of cementing.

Have your mud balance standing by.

The proportion of water to cement? Well, there are arguments there, but there is a good average figure and that is 6.3 US gallons per 94lb sack of cement. If you are using a bentonite slurry then it will be 7.0 US gallons per 94lb sack. You will have to convert this yourself, or see the attached conversion table (figure 5-106A) which will help you out.

Remember — what has been said, and will be said, is a guide to help you get started on methodology.

Why do you need the mud balance? Because when that cement slurry comes out at about 14.5 pounds per gallon (1.74 SG), you will be ready to pump.

Have your well trained crew standing by, check all is in order, and give the command.

When you have finished pumping, immediately flush the pump — with the fresh water you had put to hand — all lines etc., then strip and clean.

When the cementing is finished, give it at least twenty-four hours to set (dependent upon volume of cement pump — sometimes for large volumes as long as forty-eight hours) during which time you would have cleaned up everything, including your next run of tools. If there was temporary casing in the hole then you would have removed it while the cement was still wet — wouldn't you!

To check the condition or hardness of the cement after the drying period, give the plug inside the casing a little "prod" when you run your tools in (no circulation). If it is still soft, leave it. If it feels hard (the cement that is) continue to check as you drill (gently does it) if it gets soft, stop and wait.

Unit Conversion Table

Fig 5-106A

	m	km	in	ft	yd	mile
length	1	0.0010	39.370	3.2808	1.0936	0.0006
	1000.0	1	39370	3280.8	1093.6	0.6214
	0.0254	—	1	0.0833	0.0278	—
	0.3048	0.0003	12.000	1	0.3333	0.0002
	0.9144	0.0009	36.000	3.0000	1	0.0006
	1609.3	1.6093	63360	5280.0	1760.0	1

	mile²	km²	ha	acre	m²	ft²
area	1	2.5900	259.00	640.00	2.5900×10^7	2.7878×10^7
	0.3861	1	100.00	247.11	1.0000×10^6	1.0764×10^7
	3.8610×10^{-3}	0.0100	1	2.4711	1.0000×10^4	1.0764×10^5
	1.5625×10^{-3}	4.0469×10^{-3}	0.4047	1	4046.9	43560
	3.8610×10^{-7}	1.0000×10^{-6}	1.0000×10^{-4}	2.4711×10^{-4}	1	10.764
	3.5800×10^{-8}	9.2903×10^{-8}	9.2903×10^{-6}	2.2957×10^{-5}	9.2903×10^{-2}	1

	m³	l	in³	ft³	gal (imp)	gal (us)
volume	1	1000.0	61024	35.315	219.97	264.17
	0.0010	1	61.024	0.0353	0.2200	0.2642
	—	0.0164	1	0.0006	0.0036	0.0043
	0.0283	28.317	1728.0	1	6.2288	7.4805
	0.0045	4.5461	277.42	0.1605	1	1.2009
	0.0038	3.7854	231.00	0.1337	0.8327	1

	g	kg	t	carat	oz	lb
weight (mass)	1	0.0010	—	5.0000	0.0353	0.0022
	1000.0	1	0.0010	5000.0	35.274	2.2046
	1.0×10^6	1000.0	1	—	35274	2204.6
	0.2000	0.0002	—	1	0.0071	0.0004
	28.349	0.0283	—	141.75	1	0.0625
	453.59	0.4536	0.0005	2268.0	16.000	1

	m/s	m/min	ft/s	ft/min	km/h	mile/h
velocity	1	60.000	3.2808	196.85	3.6000	2.2369
	0.0167	1	0.0547	3.2808	0.0600	0.0373
	0.3048	18.288	1	60.000	1.0973	0.6818
	0.0051	0.3048	0.0167	1	0.0183	0.0114
	0.2778	16.667	0.9113	54.681	1	0.6214
	0.4470	26.822	1.4667	88.000	1.6093	1

	kg/m³	kg/l=g/cm³=SG	lb/in³	lb/ft³	lb/gal (imp)	lb/gal (us)
density	1	0.0010	—	0.0624	0.0100	0.0083
	1000.0	1	0.0361	62.428	10.022	8.3452
	27680	27.680	1	1728.0	277.41	230.99
	16.018	0.0160	0.0006	1	0.1605	0.1337
	99.779	0.0998	0.0036	6.2290	1	0.8327
	119.83	0.1198	0.0043	7.4807	1.2009	1

	kN/m²	bar	atm	kp/cm²	lbf/in² (psi)	ft w.g.
pressure (stress)	1	0.0100	0.0099	0.0102	0.1450	0.3346
	100.00	1	0.9869	1.0197	14.504	33.458
	101.32	1.0132	1	1.0332	14.696	33.901
	98.066	0.9807	0.9678	1	14.223	32.811
	6.8948	0.0689	0.0680	0.0703	1	2.3068
	2.9888	0.0299	0.0295	0.0305	0.4335	1

	kW=kNm/s	W=Nm/s	kpm/s	hp	ft lbf/s	
power	1	1000.0	101.97	1.3410	737.56	
	0.0010	1	0.1020	0.0013	0.7376	
	0.0098	9.8067	1	0.0132	7.2330	
	0.7457	745.70	76.040	1	550.00	
	0.0014	1.3558	0.1383	0.0018	1	

	kNm	Nm	kpm	in lbf	ft lbf	
torque	1	1000.0	101.97	8850.8	737.56	
	0.0010	1	0.1020	8.8508	0.7376	
	0.0098	9.8067	1	86.796	7.2330	
	0.0001	0.1130	0.0115	1	0.0833	
	0.0014	1.3558	0.1383	12.000	1	

length	**area**	**weight (mass)**	**force/work**	**pressure/stress**	**m**—milli (10^{-3})
cm—centimetre	**ha**—hectare	**g**—gramme	**kp**—kilopond	**atm**—atmosphere	**c**—centi (10^{-2})
m—metre	**acre**—acre	**kg**—kilogramme	**lbf**—pound force	**bar**—bar	**k**—kilo (10^3)
km—kilometre		**oz**—ounce	**kpm**—kilopond metre force	**w.g.**—water gauge	**M**—mega (10^6)
in—inch		**lb**—pound	**ft lbf**—foot pound force		
ft—foot		**t**—tonne	**N**—newton		
yd—yard					
mile—mile		**volume**	**power**		**/**—per
		gal (imp)—Imperial gallon	**W**—watt		**s**—second
		gal (us)—United States gallon	**hp**—horse power		**min**—minute
		l—litre		**s g**—specific gravity	**h**—hour

106

Cementing methods

We will try to take these from the least difficulty upwards which, we feel, follows from the shallowest downwards.

Hand pouring

For shallow open holes, or for shallow holes with casing and a nice big annulus, this is still a good, simple method provided you use something like the bentonite or polymer additives. Shallow? Not more than one hundred feet.

The mixture will be of the same proportions as above. It still has to have a similar slurry weight and it should displace any water or light muds (you shouldn't be using anything heavy at this depth) that are there.

If you have a good crew, good casing and a touch of expertise you can use this to considerable advantage — but only at shallow depths.

For casing a shallow hole, get your casing ready with the "shoe" of the first joint blocked with cement — known as a drillable plug. Mix and pour the required amount (plus a bit for fissures etc.) of cement into the open hole then run the casing in.

The casing with plug will displace the cement slurry into the annulus. You might have to add a bit of weight to effect displacement, by filling the casing with water or mud or even by a slight push from the top — make sure you are using centralisers.

Bridging

If, for some reason, the "top hole" which needs cementing overlays a hole of a smaller diameter that doesn't, then make up a cement plug similar in shape to that shown in figure 5-107A (shallow holes remember) which allows it to "run". It should be slightly less in diameter than the top hole on its biggest diameter, and slightly less than the lower hole on the smallest.

Drop this plug into the hole, smallest diameter first, and it should settle at the change in diameter of hole and bridge the gap — a sort of a "bridge plug". Cement on to the plug, wait for it to dry, then drill it out.

If you have an open hole where only the top needs cementing, then calculate the volume of hole *not* to be cemented, put that volume of fine sand (needs to be fine to prevent infiltration of cement) into the hole and cement on top of the sand — then drill out.

If you have a hole for casing when only the top of the casing needs cementing then, at the required point, fit a "shale trap" (figure 5-108A) or similar, to your casing as it is lowered, pour the cement, and the shale trap will hold it.

Fig 5-107A

SIMPLE BRIDGE PLUG/CEMENTING PLUG

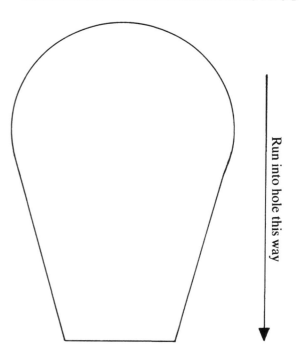

Run into hole this way

107

SHALE TRAP

Fig 5-108A

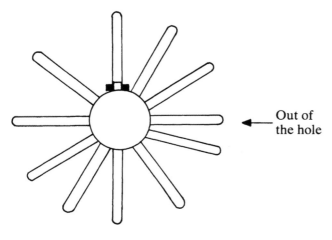

Out of
the hole

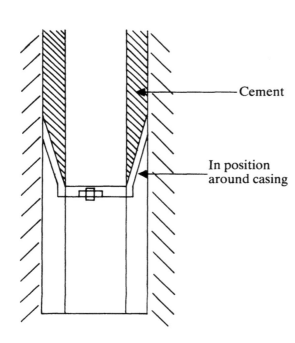

Cement

In position
around casing

Tremie pipe

For slightly deeper holes, say, from surface to not more than 200 feet (you are not constrained to pouring cement by hand for the first 100 feet or so) conditions and available equipment will dictate the method to be used, but a "tremie" is a good thing to consider.

What is a tremie? It is a long rigid pipe (or pipes screwed together) which is connected to your cementing pump by a hose, through which you pump cement.

There are two uses for the tremie — which we like to see have an inside diameter of 1–1½" — and they are inside and outside the casing.

To work outside the casing you must have an adequate annulus through which to pass the tremie unheeded and get it out again.

Run your casing, which should have a drillable plug in the shoe, and get the weight of your rig on the casing to stop it bobbing up; filling with water is sometimes good enough.

Run the tremie down the annulus to about twelve inches from bottom, pump the required amount of cement and then, and only then, pull the tremie back above the cement. Pump clean water through it to clean it out as you retrieve. This clean water is not sufficient to fully clean the pump — that has to be stripped and cleaned.

Your cement pump, if different from your mud pump, should have a pressure rating commensurate with the pressure needed (and more) to overcome frictional losses inside such a small tremie and resistance against a head of water.

This method is handy for a "top and bottom" job where the client only wants the bottom few feet of the casing cemented, then backfilling, then the top few feet.

The tremie inside the casing is a different kettle of fish. It is a little more complicated, but vastly rewarding, because you get a positive cement job which is usable to much greater depths.

The "tremie" in this case can be drill-pipe provided you are using internal flush drill-pipe, (that is, drill-pipe which is smooth inside bore all the way — remember?). Drill-pipe that is internally upset could cause an obstruction to the flow of cement slurry, and deposits of cement could be left inside the drill-pipe after washing, with disastrous results if they come loose when you are drilling with, say, a hammer.

Anyway — the whole operation centres around *the* drillable plug in the shoe. There is a hole in the plug (figure 5-109A) and set in that hole is a plastic (drillable?) non-return valve — these whole cementing shoes, including plug, valve, shoe etc., can be purchased as a proprietary item or you can do it yourself.

Fig 5-109A

CEMENTING SHOE WITH N.R.V.

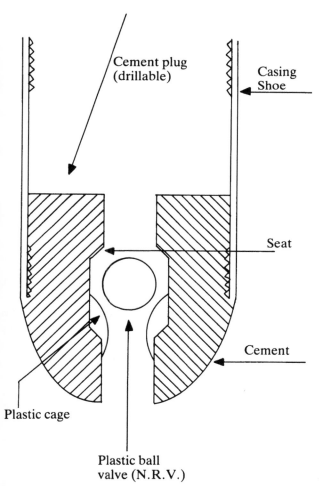

Cement plug (drillable)

Casing Shoe

Seat

Cement

Plastic cage

Plastic ball valve (N.R.V.)

Run your casing, with the special "shoe" attached, until it is a couple of feet off bottom. Then the tremie or drill-pipe is located in the hole in the shoe in the region of the valve and cement is pumped. The cement can't come back inside the casing because of the non-return valve so, when the required amount of cement has been pumped, and if the casing is at surface you will see the cement coming out of the annulus (give it a minute or two to flow if you have cement left over), you "back-off" the tremie and flush out with clean water as you trip. Then you strip and wash the pump.

When the cement is dry you just drill the shoe out and the valve will come to surface in a less than mint condition.

There will be times when, even at 30% extra slurry, the cement will not reach surface due to some unexpectedly large fissures. Then you wait until the cement is dry, probe the annulus — probably with the tremie pipe — find the cement level, mix another batch, and cement down the annulus.

That special shoe is known as a "float shoe" or, if manufactured by the Baker Tool Co., a "Baker Float". We do not normally mention manufacturers' names but in this case the internationally accepted name for this shoe is "The Baker float" — good people they are too.

The "Baker float", or similar, has another function. If you are running heavy strings of casing (remember what we said about running drill strings with a bit non-return valve?) you can "float" them down in the mud or water in the hole, thus saving wear and tear on the rig — you might even have to put water inside the casing to help it down.

Displacement method

This is probably the most widely used system of all for cementing the deeper holes and this is where you need the cementing cap on top of the casing to be cemented.

Study very carefully figure 5-110A to see how the cap is constructed — it should last you a very long time if looked after. The cap screws into the top of the casing and from it comes an "elbow" terminating in an on/off valve, both at least the

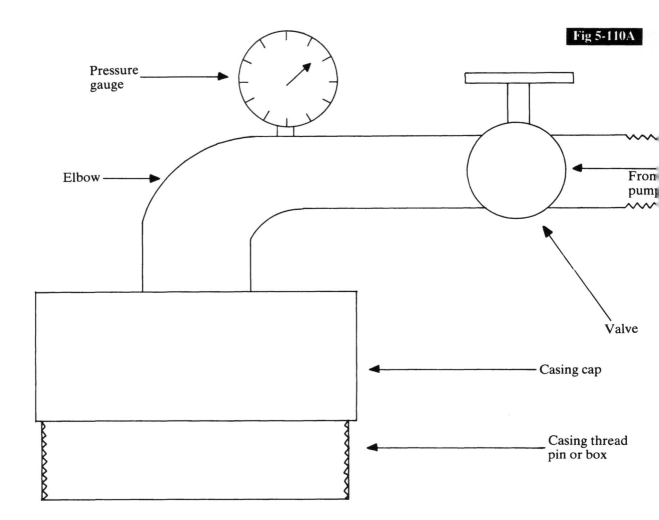

Fig 5-110A

Pressure gauge

Elbow

From pump

Valve

Casing cap

Casing thread pin or box

same inside diameter as the cementing line from your pump (at least). There is also an all important pressure gauge in the "elbow".

You will also need a pig. Not a live one, merely a drillable plug similar in shape to the "bridge plug" used for top hole cementing — see figure 5-107A. The "pig" is slightly less in diameter than the ID of the casing to be cemented and tapers down a little at its lower end — this is to prevent it bridging when it goes down the casing.

Here is what you do:-

1) Install the casing with centralisers (open ended) and clamp tight in the rig, leaving the shoe a foot or two off bottom.

2) Fit the casing cap and hook up the cementing lines.

3) Mix the requisite amount of cement and have a pit (or pits) of mud at the ready. Also have a

good supply of fresh water close to the pump suction.

4) When the cement has reached a satisfactory condition, pump it inside the casing via the casing cap.

5) Transfer the pump suction into the mud pit.

6) Look at the pressure gauge on the casing cap and if OK remove the casing cap from the casing, drop in the "pig" (the right way up) and replace the casing cap and lines.

7) Pump mud. This will push the pig down the casing which in turn will force the cement up the annulus.

8) When the cement appears on surface (or almost all the cement has gone if the "mouth" of the casing is below surface) give the cement a "squeeze" by momentarily running the cement pump fast thus building up pressure.

9) Turn off the valve in the casing cap elbow, thus preventing any backflow of mud etc., and remove the cementing line.

10) Transfer the pump suction into the clean water and flush out the pump and cementing lines, directing the line away from the mud pits, don't forget.

11) Strip the pump and give it a good clean.

From then on, just wait the required time for the cement to go off and check that there is (or isn't) any pressure in the casing from the casing cap pressure gauge. If there is, gradually open the valve to discharge the pressure — making sure nobody is in the way — then remove the cap. No pressure, remove the cap.

Do your normal probing to test the cement, drill out the plug if all is well, then carry on. A nice, neat job — eh?

In deep wells a combination of the cementing shoe with NRV and displacement with a pig is normal.

Cement pumps

Almost anything will do that is equal to the work in hand. If your cementing work is small then you could conceivably use your foam pump. If you have a lot of cementing then use your mud pump — after all, you are going to dismantle and clean them immediately after every cementing job, aren't you!

If you are rich enough to have a separate pump then buy one, just bearing in mind that you want to get that cement in the hole as quickly as possible and that the pump pressure is sufficient to overcome all backpressures and frictional losses.

We favour a triplex pump because — size for size you get a better "package" than with all but the centrifugal and we don't favour the latter because it is not positive enough with these high specific gravities — at least, we don't think so.

Ideally, the cement should be fed directly into the pump suction (gravity fed) on the same level, or, in major works, a separate pump, on the same level as the cement (gravity fed), pumps into the suction of the main pump.

Always take advice!

6 Well Development

We have been informed of a case of a major water well contract in Africa where the contractors used bentonite for their drilling in a haphazard way and failed to develop the wells correctly after completion.

The client is still visiting finished well-sites long after the contractor had gone, cleaning out flow meters which are choked with bentonite and cuttings.

That surely is enough to introduce this next lesson and the importance which must be given to the development of water wells.

The first thing to remember is that *all water wells have to be developed* whether in soft or hard formations.

Read on.

Well Development

If you have been using bentonite to drill, you have got to get rid of that awful cake on the walls of the hole and attempt to remove bentonite that has infiltrated the acquifer(s). For this you must use a chemical which must be harmless — there are some that do the job but are poisons. A couple of those usable are Tetra Sodium Pyrophosphate and Sodium Hexametaphosphate — there are also proprietary names such as Kalgon (Calgon) and Barafos. Consult your friendly mud company and it will help in this matter. A collective name — Polyphosphates.

Remember, it is no use just mixing up a bunch of these polyphosphates, pouring the slurry into the hole, and leaving the process to work — it doesn't. What follows will show you how to go about this essential task but, as regards mixtures, follow the manufacturers' recommendations, although an average mix would be 4–6lbs per 100 gals. of water.

Good development can turn an ordinary well into a good one.

The difference between the static water level (also known as rest water level) and dynamic water level is drawdown. In other words, when you are not pumping the well, the water will "rest" (static) at a certain level and when being pumped the water will draw down to the dynamic level.

When the pump is shut off the water will return to static (rest) level which is known as "recovery".

Let us first look at some of the tools, old and new, used for well development:-

Bailer

Used mainly by the cable tool brigade it is exactly as described earlier in this series of lessons and is illustrated in figure 3-68A. The function is to give your bailer a nice fit inside the casing. (Not too tight). This is lowered down into the sump below the screens and as it is withdrawn it will create suction underneath itself. As it passes the screens it will tend to draw whatever is outside the screen such as fines too small for the screen — inwards.

As the bailer is lowered again it will tend to push outwards through the screen. After many such cycles, during which sand that has settled into the sump will be drawn into the bailer through its flap valve, the sands, gravels etc. outside the screen will gradually settle themselves into a nice, graded, natural gravel pack.

So you see how important this method is when using stainless steel wire wound screen. But don't overdo the speed of the cycle because, firstly, you won't give the bailer time to go all the way down, and secondly, you could collapse the screen.

When the bailer is returned to surface, which you would do from time to time, you would empty the sand from inside. A variation of the "common bailer" is the "sand pump", which is more effective in drawing the sand into itself (figure 6-114A)

This agitation around the screens can do wonders for mud residue if the water in the hole has some polyphosphates mixed in.

Run the bailer on your sand line (remember?) and, to repeat, don't have it too close a fit in the casing as, with sand about, it could become tight in the casing.

Surge blocks

There are two types of these — solid, and valved, as shown in figure 6-115A.

They are, as you see, sandwiches of wood and rubber, the rubber being the seal. Again — not too tight in the casing when there is sand about, it could lock in.

Do not use these blocks inside screens, because they create quite a bit of suction to do the job already described under "bailer" (settling the natural pack outside the screen). They are best used 10–15' below the static water level.

The valved surge block has less "surge" on the downward stroke because some of the water passes up through the valve, but it is more gentle on the screens and is preferred. These would normally be used on the end of drill-pipe on rotary rigs, or under the drilling stem for cable tool.

You will need to "stroke" surge blocks about three to four feet up and down, but keeping them well away from screens. After some time remove the block from the hole, run in your bailer or sand

Sand Pumps Wire Line and Rod Operated

Fig 6-114A

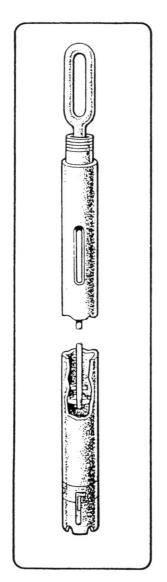

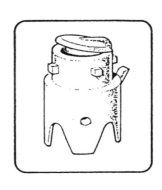

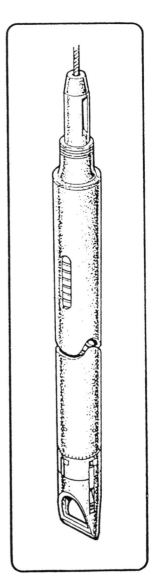

pump, clear sand from the hole bottom, measuring the quantity with each cycle (that way you know how development is progressing) and so on, and so on, until there is no more sand or bentonite (you have polyphosphates in there haven't you!).

With the valved type block you could eventually see water coming to surface because water will pass through the valve on the

downward (surge) stroke and be retained on the upstroke.

Developing with compressed air

There are two methods here and both are variations on the theme of air lift pumping. Therefore — to be effective — the submergence

114

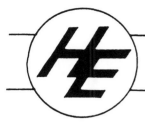

Waterwell Development and Testing

Prepared for Hands-England Drilling Ltd by
Bridge Ironworks (Engineering) Ltd
waterwell engineers and consultants

Fig 6-115A

The development and testing of a waterwell has to be performed after the well has been drilled in order to establish its maximum potential capacity. To this end a systematic programme of work has to be undertaken to clean out and open up the water-bearing formations around the well, to permit the free flow of clean water into the well, and to test the rate of flow which the well can yield.

Well Development

The process of well construction will nearly always result in a deterioration in the local flow capabilities of the water-bearing formation.

The purpose of well development is twofold: firstly, to remove the effects of drilling (i.e. the mud cake on the walls of the hole, the silt which has penetrated the formation and any compaction wich may have occurred) and, secondly, to increase the formation permeability by the removal of fine particles naturally present. The latter is especially the case in unconsolidated aquifers where the end product of development is a natural filter medium of the coarser particles around the well.

Simple pumping of a completed well will result in little or no development of flow rate since "bridges" of coarse and fine material are soon established effectively, reducing permeability. What is required is repeated reversals in flow direction. On the outflow the stability of the bridges is destroyed and on the inflow the finer particles are pulled into the well. These must then be removed by bailing or pumping at regular intervals (fig. 1).

DESTRUCTION OF BRIDGES OF SAND GRAINS

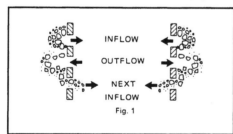

INFLOW

OUTFLOW

NEXT INFLOW

Fig. 1

This phenomena can be achieved in a number of ways:

1. By alternate pumping and recharging –
   (a) Using bailer and recharge hose alternately;
   (b) With submersible pump (without foot valve), allowing the water in the riser pipe to flow back at regular intervals.

2. By use of surge plungers (fig. 2).
   These are operated up and down in the well casing on weighted line about 10 feet below S.W.L. The operation should be carried out slowly at first to avoid collapsing the well screen with the vacuum created on the up-stroke, and should be interposed with regular bailing to remove accumulated sand.

VALVE TYPE SURGE PLUNGER

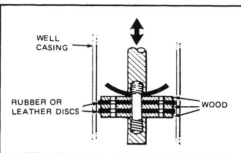

WELL CASING

RUBBER OR LEATHER DISCS

WOOD

a) OPERATED ON DRILL STEM BELOW WATER LEVEL

b) VALVE CAN BE OMITTED, IN WHICH CASE BLOCK IS SOLID WITH JUST TWO DISCS AND WILL GIVE A MORE VIGOROUS SURGING ACTION.

Fig. 2

Hands-England DRILLMASTER A120
Foam-flush drilling - Libya

HANDS-ENGLAND DRILLING LTD · LETCHWORTH · HERTS · ENGLAND

Lift in feet	50			75			100			125			150			175			200			250			300		
Length of Main in feet	150			206			275			333			375			403			440			500			570		
Submergence Ratio	2 : 1			1.75 : 1			1.75 : 1			1.66 : 1			1.5 : 1			1.3 : 1			1.2 : 1			1 : 1			0.9 : 1		
Air/Water Ratio	2 : 1			3.1 : 1			4 : 1			4.7 : 1			5.2 : 1			6.25 : 1			7.5 : 1			9 : 1			12 : 1		
Air Press: Theor	43 lb/sq in			56 lb/sq in			76 lb/sq in			90 lb/sq in			97 lb/sq in			99 lb/sq in			104 lb/sq in			108 lb/sq in			117 lb/sq in		
Volume of Water Gals per hour (Imperial)	Dia of Rising Main (in)	Dia of Air Main (in)	Volume of Air Cu ft/min	Dia of Rising Main (in)	Dia of Air Main (in)	Volume of Air Cu ft/min	Dia of Rising Main (in)	Dia of Air Main (in)	Volume of Air Cu ft/min	Dia of Rising Main (in)	Dia of Air Main (in)	Volume of Air Cu ft/min	Dia of Rising Main (in)	Dia of Air Main (in)	Volume of Air Cu ft/min	Dia of Rising Main (in)	Dia of Air Main (in)	Volume of Air Cu ft/min	Dia of Rising Main (in)	Dia of Air Main (in)	Volume of Air Cu ft/min	Dia of Rising Main (in)	Dia of Air Main (in)	Volume of Air Cu ft/min	Dia of Rising Main (in)	Dia of Air Main (in)	Volume of Air Cu ft/min
1,000	1½	½	7.0	1¼	½	8.25	1¼	¾	10.6	1¼	¾	12.5	1¼	¾	15.0	1½	¾	16.6	1½	¾	30	2/1½	¾	30	2/1½	¾	40
1,500	1½	½	8.0	1½	¾	12.5	1½	¾	16.0	1½	¾	18.8	1½	¾	25.0	2	¾	25.0	2/1½	¾	35	2	1	45	2½/2	1	55
2,000	1½	½	11.0	1½	¾	16.6	2	¾	21.4	2	¾	25.2	2	¾	30.0	2	1	33.0	2	1	45	2½/2	1¼	55	2½/2	1¼	70
3,000	2	¾	16.0	2	¾	24.8	2	¾	32.0	2½	1	37.5	2½	1	45.0	2½	1	50.0	2½	1	65	2½	1¼	80	3/2½	1¼	96
4,000	2½	¾	22.0	2	1	33.0	2½	1	42.6	2½	1	50.2	2½	1	60.0	2½	1¼	67.0	3	1¼	80	3½/3	1¼	100	3½/3	1¼	128
5,000	2½	¾	26.6	2½	1	41.0	3	1	53.3	3	1	62.7	3	1¼	75.0	3	1¼	83.0	3½/3	1¼	100	3½/3	1¼	120	3½/3	1½	160
6,000	3	1	32.0	3	1	49.6	3	1	64.0	3	1¼	75.2	3½	1¼	90.0	3½	1¼	100	3½/3	1¼	120	3½/3	1½	144	4/3½	1½	192
8,000	3½	1	44.0	3½	1	68.7	3½	1¼	88.8	3½	1¼	104.0	4	1¼	115.0	4	1½	139	4/3½	1½	168	4/3½	1½	200	4½/4	2	266
10,000	4	1	53.2	4	1¼	82.5	4	1¼	106.0	4	1¼	125.0	4	1½	138.0	5	1½	166	4½/4	2	200	4½/4	2	240	5/4	2	320
12,000	4	1¼	64.0	4	1¼	99.4	4	1¼	128.0	5	1½	151.0	5	1½	166.0	5	1½	200	5/4½	2	240	5/4½	2	290	5/4½	2	360
15,000	5	1¼	79.8	5	1½	125.0	5	1½	166.0	5	1½	188.0	5	1½	208.0	6	2	250	6/5	2	300	6/5	2	360	6/5	2½	450
20,000	5	1¼	106.4	5	1½	166.0	5	1½	214.0	6	2	252.0	6	2	277.0	6	2	333	6/5	2	400	6/5	2½	480	7/6	2½	600
25,000	5	1½	133.0	6	2	206.0	6	2	266.0	6	2	312.0	6	2	346.0	7	2	415	7/6	2½	500	7/6	2½	600	7/6	3	750
30,000	6	1½	159.6	7	2	248.0	7	2	322.0	7	2	375.0	7	2	420.0	7	2½	505	8/7	2½	585	8/7	2½	685	8/7	3	870
40,000	7	2	212.8	8	2	332.0	7	2	426.0	8	2½	502.0	8	2½	535.0	8	2½	670	8½/8	2½	770	8½/8	3	920	9/8	3½	1150
50,000	8	2	266.0	8	2½	412.0	8	2½	533.0	8	2½	627.0	8½	2½	666.0	9	3	830	10/9	3	970	10/9	3	1130	10/9	3½	1440
60,000	8½	2	319.2	8½	2½	496.0	8½	2½	640.0	9	2½	752.0	9	3	800.0	10	3	1000	10/9	3	1170	10/9	3½	1360	11/10	3½	1750

Fig 6-116A

of the system in water relative to what is above the water is all important. Let us, therefore, look at the principles of air lift first.

Firstly look at figure 6-116A which illustrates a basic 2-1 submergence level which, at the depths shown, will work nicely if the amount of compressed air and sizes of pipes shown are used. Then carefully study the submergence relative to depth, volume of water, sizes of pipe and quantity of air.

For example, if you have to lift the water 100 feet (above dynamic water level) from a well which is producing 3000 Imperial gallons per hour, the length of the rising main would have to be 275 feet, which is a submergence of 1.75-1. The rising main should be 2″ inside diameter. The air pipe (a little longer than the rising main — see later) should be 3/4″, air pressure 76 psi and air volume 32 cfm.

If you cannot satisfy the submergence then air lift will not work very effectively, although by careful control of air flows some success is possible. Anyway, let us look at what to do. And don't forget the polyphosphates.

The first method is the simplest but needs a lot of experience to get it going; you will duplicate the rising main and air pipe with your casing and drill-pipe. In other words run your drill-pipe into the hole "open-ended" (without a bit or hammer etc.) until the mouth of the pipe is about two feet above the top of the screen then inject a small amount of compressed air and wait.

The first thing you will see is the mud column flow over the casing (the wall cake is still there don't forget) and it will gradually accelerate (the time taken to do so will depend upon the amount of air — don't use too much — see air lift tables) until it shoots into the air. Then shut the air valve.

Repeat this cycle until the water appears to be clean then continue until the water *is* clean. What you are doing is to slightly pressurise the screens,

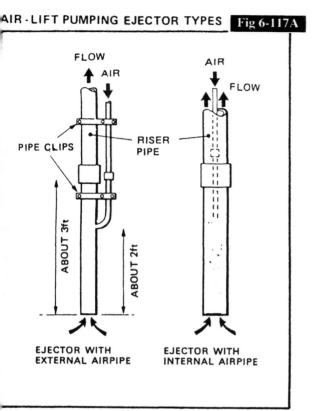

FLOW
AIR

AIR
FLOW

PIPE CLIPS

RISER PIPE

ABOUT 3ft

ABOUT 2ft

EJECTOR WITH
EXTERNAL AIRPIPE

EJECTOR WITH
INTERNAL AIRPIPE

forcing outwards as you start to lift the column. As the column accelerates up the hole you draw from the screen, thus settling and grading the natural pack. Also, water falling back into the hole when the valve is shut will will push outwards into the aquifer.

We must confess that this method can be a little slow and ponderous, but is handy to know about if submergence levels do not allow a full air lift.

To develop by air lift with good submergence you need to set up the system, matching your tools to the expected yield of water from the hole, referring to the table in figure 6-116A.

You will have a rising main, on top of which will be an elbow and an extension pipe therefrom. The elbow will have an hole in it to allow the passage of the air pipe into and down the riser. You should have a quick opening valve — handy to the operator — breaking the line from the compressor into the top of the air pipe. The air pipe must be longer than the riser.

Place the rising main in the hole 2–3' above the bottom of the screens fixing the elbow and extension pipe on surface. Run the air pipe through the elbow: you will need a pipe wiper of sorts here because you are going to move the air pipe up and down, probably on your winch, and you will have to allow for the coupling to go through. Connect the air line to the air pipe.

With the air pipe about two feet above the mouth of the riser, turn on a small amount of air (see table in figure 6-117A) and pumping will start. Wait until the mud has gone and you are getting clear water and then shut off the air. Run the air pipe a couple of feet below the mouth of the riser and quickly turn on the air fully for a few seconds. Turn off the air, pull the air pipe back into the pumping position and pump, again using a little air.

Repeat this cycle many times, gradually pulling the riser back up the screens until all the screens have been treated and the water is clean and free of solids.

What you are doing is settling and grading the natural pack around the screens, as already discussed above.

In the old days you had to have an air receiver between the compressor and the air lift tools to effect the sudden surge (air fully on) but modern compressors take care of that.

When pumping during development, have a vessel of known volume standing by, such as a 5 gal. container, and a stop watch. Time how long it takes to fill the vessel from the discharge and you can estimate how development is going — the time should be shorter as development gets better. You can also estimate the yield from the well.

Before going on, let us have a word about getting rid of clay based (bentonite) muds from the aquifer(s). If you have to use them you've got to get rid of them to make the well yield and we have already told you how. It takes time — a lot of time — and time is money. Polymer muds do not require this lengthy action, which can be days and days and can even cost you the well, due to complete blockage.

Whatever system of development you use, having drilled with bentonite you must use polyphosphates to get rid of it.

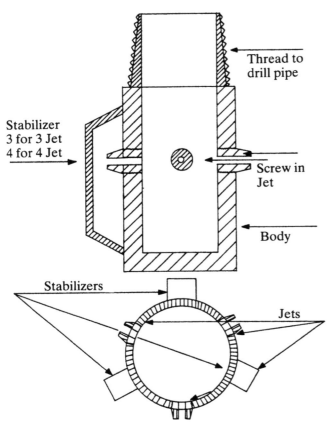

Fig 6-118A

JETTING TOOL

Thread to drill pipe

Stabilizer
3 for 3 Jet
4 for 4 Jet

Screw in Jet

Body

Stabilizers

Jets

Figure 6-118A illustrates just such an item, which can be used with air or water subject to "nozzle" sizes; can be complicated or simple; and is without a doubt very effective indeed. The jetting action is concentrated over a small area of the screen at any one time, thus "re-arranging" the natural pack and washing out mud very nicely. It is passed up and down the screen, rotating slowly, thus covering the whole screen.

Rotary rigs would run this tool on the end of drill-pipe. The simple set-up would be something like a blank ended sub screwed onto the pipe with 2–4 holes drilled into it horizontally. The more complicated jetting tool would have a series of replaceable jets of different sizes for different applications. Three or four stabilising webs should be fixed to the tool to keep it central in the casing, but construct the body of the tool fairly close to the inside diameter of the screen.

Cable tools would have the jetting tool on the end of good quality coupled tubing, and on the top end a connection from their water pump or compressor.

Whatever type of rig, the addition of polyphosphates to the water or air is very effective indeed.

The holes should be equally spaced around the tool. For instance, if there were only two they would be opposite each other; three — at 120 degrees; four — at ninety degrees. The size and number of holes are governed by the size of your mud pump or compressor. The diameter should be such as to give a minimum water/air velocity of 150 feet per second (max 300 fps) — see figure 6-119A for velocities and volumes — and the number of holes (jets) to give you enough volume at the minimum velocity to enable gentle flushing of the hole at the same time as development is taking place, thereby removing solids.

When using a jetting tool with air, please remember the rule that 2.31 feet head of water is equivalent to 1 psi of pressure. This must be taken into consideration in getting the tool started. For instance, if the bottom of your lowest screen is under a 235 feet head of water and you only have 100 psi at your compressor, even disallowing friction losses the tool will not work because water pressure at that depth is about

Fig 6-119A

JETTING TOOL — APPROX VELOCITIES & DISCHARGES

Nozzle Size in Inches	Pressure 150 PSI		Pressure 200 PSI		Pressure 250 PSI	
	Velocity Feet Per Sec	Discharge US GPM	Velocity Feet Per Sec	Discharge US GPM	Velocity Feet Per Sec	Discharge US GPM
3/16	150	12	170	13	190	15
1/4	150	21	170	23	190	26
3/8	150	46	170	53	190	59
1/2	150	82	170	93	190	104

102 psi and your 100 psi won't start it. But if you have 150 psi under that head of water then you will disperse the head very rapidly and all is well.

With air jetting, as with water, it is necessary to wash out any mud that might be there, so connect in your foam pump and inject a mixture of water and polyphosphates into the air stream, and away you go.

With water jetting, mix the polyphosphates in water in your now clean mud pits and gently circulate as you develop, thus removing unwanted solids.

The application is very simple. You slowly rotate the tool (cable tool hands will have to do it by hand) and, equally slowly, move it up and down over the entire length of your screen(s), thus applying the jet to all formations through the screens. You will be very surprised how quickly the natural pack will be settled and the water cleaned.

Gravel packed wells

So far we have only talked about naturally packed wells behind stainless steel wire wound screens, but what about those wells where artificial gravel pack is installed?

A few dimensions to think on. For a good average gravel pack your annulus around the casing/screens should be about 65mm–75mm which will give good "pouring" characteristics and a nice size for settlement of the gravel. The gravel must be rounded and NOT quarry type chippings and, say, for a screen slot size not exceeding 0.5mm the size of the gravel would be sized at passing 3mm mesh but not passing 1mm mesh. If gravel is too large or the annular space too small — or both — then there will be a tendency for bridging causing voids, or worse, no gravel at all at depth.

We are back to bentonite again because, assuming the correct grade and shape of gravel has been installed, you have still got to get rid of that nasty layer of bentonite cake that is sandwiched between the gravel pack and the wall of the hole, as well as settling the pack. If you used polymer mud you wouldn't have this problem.

Surge blocks might re-arrange the gravel to a usable pack, but it will have little effect on the mud cake even when polyphosphates are introduced. So, what do you do?

1. You can air lift and introduce polyphosphates at the same time — in effect reverse circulate by air lift. This helps.
2. You can use your jetting tool and polyphosphates and be patient. The problem here is that, with all but wire-wound screen, the number of openings in the screen are few, therefore the time taken for development is that much more.

Don't think you can pour a polyphosphate

slurry down there and wait for it to act — you will have to wait forever.

The use of clay based muds in water-well drilling is a potential disaster and you must give this problem a lot of thought in *advance* of doing the job — time is money.

Don't forget that a well drilled in rock still has to be developed because, even if you haven't used mud, debris disturbed by the bit, especially the hammer, will lodge in the fissures. Make yourself a nice jetting tool, hook up your foam pump and compressor — this time with a foam slurry — and give the well a good flushing. It works wonders. Don't forget you have to do this in all areas of the well.

Testing the Well

We will now assume that the hole has been developed, is producing water, is free of solids and that you want to test the quality and quantity of water.

Don't think you can give the well a quick pump then shoot off — it has to be pumped for at least twenty-four hours *continuously*.

Pumping can be done either by the air lift method already described or by a submersible pump, (if you are sure there is no sand), the installation of which is variable according to type. Therefore, follow the manufacturers' instructions. What we are concerned with here is measuring the volume.

Over the years we have seen some weird and wonderful ways of doing this, but the two methods most used are through a flow-meter or into a weir tank, we will, therefore concentrate on these.

Flow meter

A flow meter is a device which measures, digitally, the flow of water through it and is fitted into the discharge pipe. The operator stands by with a watch reading off the volume at a given time and notes it in a book, measuring drawdown at the same time — simple? Yes, but in the majority of cases we have seen, this reading has

been checked against the discharge over a weir tank, so:-

Weir tank

As you will see from figure 6-122A a "weir" system is actually two tanks, (sometimes more), the first taking the water from the hole and decelerating its flow and the second being the weir tank itself, which slows the water even more with baffles until it passes out of the system over a device from which the measurement is taken.

This "device" can be a "notch" or a "weir". Figure 6-121A gives tables for measuring water flow over a 90 degree "V" notch and 12″ and 24″ weirs. The height of water over the device is measured off in inches (and parts of an inch) by a "ruler" set in the tank *at least two feet back* from the actual point of discharge, thus avoiding a false measurement created by the natural tendency of water to "dip" before it flows over an obstruction. This measurement is related to the attached tables and, for instance, 3½″ of water over a 90 degree "V" notch represents 2,610 imperial gallons of water per hour.

Weir tanks should, wherever possible, resemble the construction shown in figure 6-122A if good accurate data are required. The 90 degree "V" notch is accurate for flows up to approx. 3,000 Imperial gallons per hour. Over that, you would be advised to consider the 12″ and 24″ rectangular weirs, as illustrated.

When the water has left the weir tank, it should be directed well away from the well-site so as not to allow the water to flow back into the hole, thus falsifying readings.

Fig 6-121A

TABLE A FLOW OF WATER OVER 90° "V" NOTCH WEIR   Imperial gallons per hour

H_w Height in Inches	0	$\frac{1}{8}$	$\frac{1}{4}$	$\frac{3}{8}$	$\frac{1}{2}$	$\frac{5}{8}$	$\frac{3}{4}$	$\frac{7}{8}$
1	110	180	200	250	310	380	460	550
2	650	750	870	990	1,130	1,270	1,430	1,600
3	1,780	1,970	2,170	2,390	2,610	2,850	3,100	3,370
4	3,650							

TABLE B FLOW OF WATER OVER 12 in RECTANGULAR WEIR   Imperial gallons per hour

H_w Height in Inches	0	$\frac{1}{8}$	$\frac{1}{4}$	$\frac{3}{8}$	$\frac{1}{2}$	$\frac{5}{8}$	$\frac{3}{4}$	$\frac{7}{8}$
0	0	85	235	427	652	905	1,182	1,483
1	1,804	2,144	2,507	2,883	3,275	3,684	4,109	4,545
2	5,001	5,468	5,946	6,439	6,940	7,461	7,987	8,527
3	9,074	9,638	10,210	10,780	11,390	11,990	12,600	13,220
4	13,850	14,490	15,150	15,810	16,470	17,150	17,830	18,520
5	19,230	19,940	20,660	21,390	22,130	22,870	23,610	24,370
6	25,140	25,910	26,690	27,490	28,290	29,080	29,880	30,710
7	31,540	32,370	33,200	34,050	34,900	35,770	36,620	37,490
8	38,390	39,260	40,150	41,050	41,950	42,860	43,770	44,700
9	45,630	46,560	47,490	48,450	49,410	50,370	51,350	52,300
10	53,270	54,250	55,260	56,220	57,230	58,220	59,240	60,260
11	61,280	62,300	63,330	64,370	65,420	66,490	67,530	68,600
12	69,650	70,710	71,740	72,830	73,940	75,020	76,140	77,250

TABLE C FLOW OF WATER OVER 24 in RECTANGULAR WEIR   Imperial gallons per hour

H_w Height in Inches	0	$\frac{1}{8}$	$\frac{1}{4}$	$\frac{3}{8}$	$\frac{1}{2}$	$\frac{5}{8}$	$\frac{3}{4}$	$\frac{7}{8}$
0	0	172	476	865	1,322	1,835	2,398	3,007
1	3,658	4,348	5,083	5,846	6,642	7,471	8,332	9,217
2	10,140	11,090	12,060	13,050	14,070	15,130	16,190	17,290
3	18,400	19,540	20,700	21,870	23,090	24,300	25,550	26,810
4	28,080	29,390	30,700	32,040	33,390	34,770	36,160	37,560
5	38,990	40,420	41,990	43,370	44,860	46,350	47,990	49,420
6	50,980	52,540	54,140	55,730	57,360	58,960	60,610	62,270
7	63,950	65,630	67,320	69,040	70,780	72,520	74,260	76,020
8	77,840	79,620	81,410	83,240	85,160	86,920	88,760	90,630
9	92,530	94,410	96,290	98,240	100,200	102,100	104,100	106,000
10	108,000	110,000	112,000	114,000	116,000	118,000	120,100	122,200
11	124,300	126,300	128,400	130,500	132,600	134,800	137,000	139,100
12	141,200	143,400	145,500	147,700	149,900	152,100	154,400	156,600

Fig 6-122A

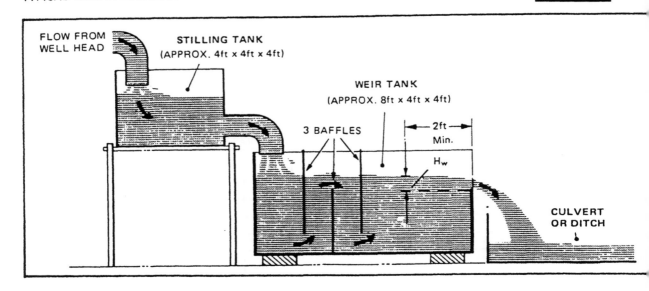

7　The Auxiliaries

Auxiliaries are those lovely things that seem totally alien to water-well drilling practices and things that the average driller only reads about. Yet some are not only tailor made for water-well drilling but are downright essential.

Management personnel should read this section and see that they can actually save money (more profit!) by spending some on such purchases as these. Some of them you don't even have to purchase — you can make them out of bits of scrap. Others you don't need immediately but need simply to have the knowledge they exist, and to know a little about them, could stand you in good stead for the future.

Is it absurd to think that, because you only drill water wells, you would never need, say, a blow out preventer or a shale shaker? Yes it is. We know of one company that actually gained a major contract because they did know.

If you can't afford all of them, then so be it but I am sure you will agree that some are essential.

Read on.

Auxiliaries

And now we begin what is probably the most important section of all for you, because each of the things we are listing has a specific purpose and is not a generalisation. Each is the culmination of what has been said before and should be studied with great concern.

A lack of knowledge of even one may make you and your operation that much less efficient. With this knowledge, and with physical experience, (remember this business is made up of experience), you will prosper.

We have visited many a site where there was a lovely new drilling machine, a magnificent mud pump, support vehicles lined up in strict order, the crew turned out like a regiment of guards, and the crew trying to mix their mud with shovels.

On another site they were attempting to drill a 5000 feet water well in a known oil and gas zone without blow-out preventers and, indeed, casing that was not cemented in.

On another the mud was so dense that you could walk on it and the crew had no thought or equipment to do anything about it; their mud pump was struggling.

Let us start then, bearing in mind that these auxiliary components are not listed in any order of priority, because they are all equally important.

Weight indicator

In previous sections we have talked about the necessity of putting a known weight on the bit (especially important with down-the-hole hammers), and yet we have seldom seen a water-well drilling machine equipped with a weight indicator, let alone on a "machine that drills". But such an addition is not at all difficult.

For a rotary table rig good examples of this type of indicator are readily available and simple to fit, usually being a "sensor" fitted into the "deadman" line in the mast leading to a gauge at the drillers position. Yes, it's as simple as that. A good system will not only measure the weight on the bit but also, with a second "needle", the actual weight of tools in the hole — essential information for the "thinking" driller.

With the wire-line "rigged" type of top-drive rig, an indicator as described above is feasible, but as most of these rigs are hydraulically powered, it is simple for the manufacturer to set gauges into a specific part of the circuit to give the same information, even if they just put a "scale" around a normal pressure gauge in the "hoist" circuit that converts pressure into weight.

For the chain-driven top-drive rig, hydraulically powered, the latter of the two methods described in the previous paragraph is recommended.

Torque gauge

This gauge is such an essential part of your work at the controls that it is almost obscene to think of life without it. The slightest variation in power needed to turn the bit will be immediately registered, and you can do something about any problem there might be. If you wait for the diesel engine to "blow coal" (a diesel will do just that under stress) you are well and truly in trouble.

To drill in lost circulation areas (see previous sections) without this gauge is unthinkable, even monstrous. By the time you know the worst it is sometimes too late.

For the top-drive rig, it is fitted in the hydraulic circuit, just like the weight indicator, but in the rotation part of the circuit, and the manufacturer does a similar thing in putting a scale around it — it costs peanuts.

For the rotary table rig which is belt or chain driven to the table, an idler pulley can be fitted into the drive which transmits a signal to a gauge at the driller's position. If the drive is hydraulic then it would be fitted in a similar way to the top-drive above — that is, around an oil pressure gauge. If you haven't got one of these then you have a "machine that drills" and not a drilling machine.

Rotational speed indicator

This shows the speed of rotation in revolutions per minute. It is very nice to have if the driller has a watch. If they don't have a watch then it is vital. With a watch you can make a mark on the drill-pipe, start drilling, and count the revs. against the time; without a watch you can't — you can only guess.

When rotary drilling, it is good to know your rotary speed to get maximum benefit in terms of performance and wear from the bit, but when using a down-the-hole hammer you *have* to know within a rev. or two what is going on or you certainly won't get maximum benefit.

Rotation speed with down-the-hole hammer and rotary drilling should be set according to the formulae in Book Nine and then adjusted to give *maximum* cutting size coming from the hole — then you have the ideal speed.

Come on you manufacturers! We have never seen a water-well drilling machine (nor, of course, a "machine that drills" — but that is to be expected) fitted with a tachometer (rev. indicator) on the rotation — mind you we have watches, *but many drillers don't*.

Pump stroke indicator

Imagine an experienced driller standing at the controls drilling away and deciding that the mud velocity looks a bit slow. Glancing across at the mud pump, the driller sees from the pump stroke indicator that the pump is already near normal maximum strokes (cycles!) and decides mud is being lost in the formations and goes about doing something to cure it.

Imagine an inexperienced driller in the same situation but without a pump stroke indicator. They would call for more and more speed, until not only will they be in trouble in the hole, but the pump will be damaged.

Now imagine that same inexperienced driller with a pump stroke indicator. They would see the pump speed, correlate it to the mud speed, come up with the right answer, and be one step nearer becoming a driller of experience.

Mud pump by-pass system

This can be simple or not so simple, but it always comes *after* the pressure relief valve in order to protect the pump.

In simple form the by-pass is a "tee" set into the

delivery line with a valve on each exit from the "tee" (figure 7-126A). The main delivery line carries on into the standpipe and the other would have a hose attached which returns into the mud pit. When you want to stop the mud — say, for changing a drill-pipe — you open the line into the pit and shut the delivery line to the standpipe by using the valves, after, of course, slowing the engine down. Mud is then "by-passed" into the pit.

It must be done in this sequence. If you shut the delivery first, both valves would be closed and the pump would be over-pressurised and the relief valve would blow, or worse.

Remember — all fittings, such as the "tee" etc., must be of sufficient pressure rating to match the pump pressure or better. Also, the main delivery line diameter should not be reduced in any way — this will restrict the flow of mud and increase pressure losses in the system.

The by-pass hose, sometimes known as a "washdown line", can be fitted with a mud gun (see later) for "gunning" (agitating) the mud, or a fitting for attaching to a jet mixer (also see later) for mixing the mud.

In less simple form you can have a number of outlets with valves, each outlet having its own job to do, such as mud gunning, jet mixing etc.

Just make sure of two things. Firstly, that your pump operator knows which valves to operate and when, and secondly that you do not starve the hole of mud by operating any of these things when drilling.

Pulsation damper

Figure 7-126B shows a typical damper as fitted to most mud pumps of the reciprocating type. What does it do? It damps out pulsations, but what pulsations? If you ever saw a mud pump without one the delivery hoses into the swivel (pump to standpipe and standpipe to swivel) would do a war dance inspired by each stroke of the pump like you never saw before, and would eventually burst with shocking results. The drill-pipe would do a nice vertical rhumba as each pump pulse blasted into the hole — replace the damper and you have relative calm. Make sure it is the correct one for your pump — they do vary in capacity.

Fig 7-126A

Fig 7-126B

126

Manometer

This is the up market word for a pressure gauge. You've got to have one on the pump as well as a mud-line pressure gauge on the rig — correlation between the two is most helpful in discovering delivery problems (and others) from the pump. We like to see it fitted where it can been seen best by the driller, and a most convenient place is on top of the pulsation damper, because not only will it be seen nicely but the pump pulsations will be damped down and the needle won't go up and down like a demented grasshopper.

Mud guns.

As already mentioned, this is a sort of a jetting tool attached to the end of a hose from the by-pass on the mud pump. When mixing mud through the jet-mixer (see later) it is a good thing to have someone holding a mud gun (with a very firm grip — they have enormous reaction) to direct a jet of mud at the point of discharge from the mixer into the pit, giving an added boost to the solution. They will, of course, gun up and down the pits as well as sorting out any "dry" bits of mud that the crew have let fall into the pits. This also boosts the overall mix.

If you have plenty of spare pump capacity or, indeed, even a second pump for mixing (some do), you can have one or more mud guns playing into the pits permanently, keeping your mud in prime condition.

Mud guns are less efficient than agitators (see later) as they will only "pierce" the top of the mud, but are certainly better than nothing.

The "sort of jetting tool" on the end of the hose is in fact a nozzle about ½–¾ of the inside diameter of the hose. The larger the hose the smaller — relatively — the nozzle, meaning, for a 2″ hose a 1″ nozzle — ½ size — for a 1″ hose a ¾″ nozzle etc.) of the inside diameter of the hose. If you think you don't need it and that the bare hose will do, get a piece of ordinary garden hose on a domestic tap and let it flow normally, then with the end of the hose squeezed in. Thats the difference — get one.

Mud agitators (figure 7-127A)

There are two main types in use in water-well drilling today. Firstly there are the mechanically driven "paddles" working in the mud suspended from a beam across the pits and powered from whatever sources are available e.g. hydraulic motors, electric motors, vee belts etc. etc. They really do work and will avoid nasty separation of the mud and water.

Do not have them running too fast — you don't want to stir up any left-over cuttings — and use them only in the suction pit, and not too near the bottom.

Oh yes — the second type of agitator is a little person with a great big shovel — take your choice.

Fig 7-127A

Mud mixer

Otherwise known as a jet-mixer, and figure 7-128A shows a general arrangement of such a unit which is very easy to make, at next to no cost, and is very efficient.

A unit of this size can handle quite large volumes of mud, certainly more than enough for most water-well operations, and don't forget, if you have a large cementing job to do, dig a third pit and mix your cement in it through the mud-mixer.

127

SCHEMATIC OF SIMPLE JET MIXER

Fig 7-128A

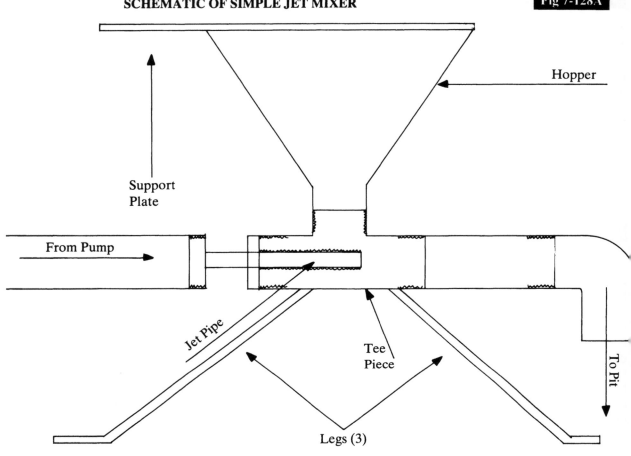

The whole thing is built around a three inch "tee" piece and a bit of one inch pipe plus a few bits and pieces of scrap.

In one end of the "tee" put a three inch male to one inch female reducer. Get the piece of one inch pipe, about six inches long, and have it threaded on the outside and screw it into the reducer until the end (inside the "tee") is about half an inch from an imaginary line drawn from the inside wall of the "vertical" part of the "tee". This is where the threading of the one inch pipe is necessary, because that half an inch is slightly variable to get the best result.

Lock the one inch pipe to the reducer with a lock-nut. On the outside end of the one inch pipe screw a fitting suitable for taking the hose coming from the mud pump by-pass system.

In the open end of the "tee" opposite to the one inch pipe (the jet) screw about a one yard length of three inch pipe with an "elbow" on the end of it discharging downwards.

The only outlet from the "tee" now is the right angled one and into that screw a nice big hopper similar to that shown in figure 7-128A with a nice big plate attached to it so that the crew can rest heavy sacks of mud whilst pouring into the hopper. Sit the whole thing on a substantial three legged stand (you can't level four legs without difficulty) and you've got yourself a mud-mixer.

Make sure the three legged stand is well spread with two legs at the back, because "every action has a reaction" and the speed of the mud going through the mixer and hitting the elbow at the

128

Fig 7-129A

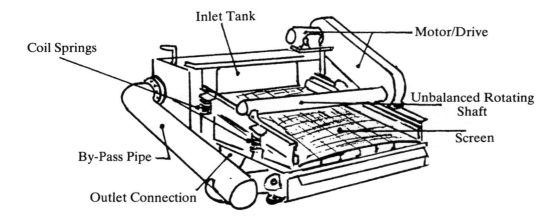

Coil Springs — Inlet Tank — Motor/Drive — Unbalanced Rotating Shaft — Screen — By-Pass Pipe — Outlet Connection

outlet will tend to force the whole thing backwards.

When hooked up to the mud-pump by-pass and with the three inch elbow facing downwards into the pit, start the pump and get some mud into the hopper. You might find you will have to adjust the "jet" a little inwards but once set it can be locked up and left. The "depression" created at the jet by the water (mud) passing through the jet will draw the dry mud nicely into the stream of water and you will have a lovely mix.

Don't forget to circulate the mud through this system for some time after the last sack has gone — "gunning" at the same time — to make the best of the mix.

Mud tanks

If you want to use those ghastly little things that the seismic "boys" use and tear your pumps and swivels apart by circulating "dirty" mud, then go ahead, but if you want to use proper tanks then your rig has got to be mounted on a substructure (see notes in book 1) high enough to allow the mud to exit from the conductor via the flow-line and into the mud tank. You can, if necessary, build a ramp of sufficient height and run your rig up on that to achieve the same result.

The mud tanks will still have to have a similar capacity as calculated for the pits, which will

make them huge. Just think in terms of moving them from site to site or, if they can be dismantled, the time taken to do that and still moving them about.

OK you have just such a rig and tanks for drilling major wells and all the mechanical handling equipment for moving then you will need a:-

Shale shaker

These can, of course, be used with any rig that has sufficient height below the working/rotary table to allow mud to flow out of the conductor via a bell-nipple and flow-line and enter the *top* of the shaker (figure 7-129A and 7-130A). You will no doubt find that your shaker has an inlet at the bottom of the "chamber", but if the flow-line is fitted there you will get a build up of cuttings around the inlet which will rapidly block off and, indeed, fill the flow-line.

What is a shale shaker?

It is a mechanically driven sieve. Mud from the flow-line enters a chamber at the back of the shaker and overflows from the chamber onto a mesh of predetermined size. The mud drops through, leaving the cuttings to be vibrated off the mesh so that they may be disposed of.

The mesh (screen) is set at an angle to the base of the shaker, therefore the shaker must be set

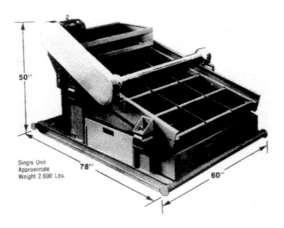

Fig 7-130A

50"

Single Unit
Approximate
Weight 2,600 Lbs.

78"

60"

So the mud is now clean. No it isn't — in fact, far from it, because it still has to find its way along the channels (or flumes) and into the pits, ending in the suction pit, dropping out as it goes. Mind you, you could make the whole thing better by following the shaker with a:-

De-sander

When using this unit in an operation where pits contain the mud, you do not have to link the settling and suction pits to allow the mud to flow from one to the other; the de-sander unit will take the mud from the settling pit, pass the mud through the de-sander and discharge it into the suction pit nice and cleanly. You see — a de-sander has its own pump which is almost always centrifugal.

When using tanks as the mud source, pass the mud from one tank to another through the de-sander.

What a splendid idea! The dirty mud comes up the hole; across a shale shaker which takes all the larger cuttings; along channels (flumes) to drop more cuttings (this time finer); into the settling pit (tank) where more drop out, through a de-sander which takes just about all that are left; into the suction pit and finally, beautifully clean, into the mud pump which is going to last you a lifetime of work because the mud is clean.

So what is a de-sander? Figure 7-127A gives a general view of a de-sander unit (including de-silters and a pair of hydrocyclones), showing the suction hose from the settling tank, the capacity of this unit is judged by the capacity of your mud system. The mud is then passed through a manifold at high velocity into one or more cones (centrifugers), the number of these cones, again, being established by the capacity of your overall system and, indeed, to the smallest size of grain to be absorbed. The bigger the system, the more cones you have, the smaller the system the fewer cones. The manufacturer should be able to help you there.

In the cones (see figure 7-131A which shows a hydrocyclone but the principle is similar) the velocity of the entering mud (the inlet is not central to the cone but tangential (offset)) imparts sufficient *centrifugal force on the mud to separate*

level to make it efficient. There can be two or sometimes more "decks" of screens, the top having the the largest holes in it, the next down smaller etc. etc. Thus the cuttings will be graded, the largest going over the top screen, the next size over the next screen down, etc.

The area of the screen — therefore the size of the shaker — is governed by the flow of mud from the hole, therefore related to the mud pump so, when ordering a shale shaker, you must give this information. The size of the holes in the screen is really governed by the material you are drilling and it is always best to take advice on this. The number of holes per square inch (or centimetre) is how a screen is measured so, 80 mesh means 80 holes per square inch. It is always advisable to have a number of different sizes of screen standing by.

By the way, do you know how a wire wound screen is designated? A 20 slot screen is not 20 slots per inch of screen; each slot is *20 thousandths of an inch* wide.

After a shale shaker, it is preferable to have the mud running along channels into the pits to get the last cuttings (fines) out of the mud. If the shaker is mounted on top of the mud tanks (yes — quite normal — see how high you have to get the rig up?). Then the mud should be discharged into the flumes.

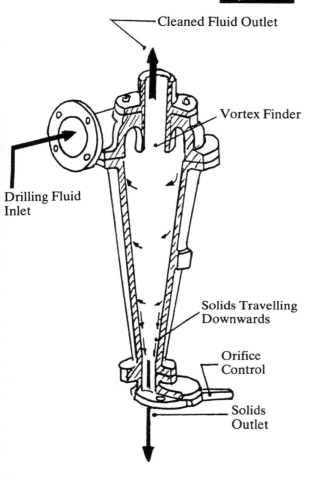

Fig 7-131A

Cleaned Fluid Outlet

Vortex Finder

Drilling Fluid Inlet

Solids Travelling Downwards

Orifice Control

Solids Outlet

the solids from the liquid — the solids gravitate out of the bottom of the cone and that lovely clean mud out of the top through a pipe, the bottom of which is slightly below the inlet, through another manifold into your suction pit (tank).

The discharge of solids (cuttings) from the bottom of the cone(s) should be almost dry and not a soggy continuous sausage; the cones can be adjusted to get the required result by altering the orifice at the bottom but, if your pump and number of cones are correct for your system, all should be well. But check anyway.

Can you see how important it is to have control over your mud and to know what it is doing? If it wasn't necessary, then people wouldn't pay all this money to do it, would they! But if you can't afford the shale shaker and de-sander you most certainly have got to have a:-

Marsh funnel and cup (figure 7-131B)

What does it do? It measures the viscosity of the mud relative to fresh, clean water determined by the time it takes one US quart to pass through the hole in the bottom of the funnel. Fresh water will take about twenty-six seconds subject to temperature, altitude, mineral content etc. but twenty-six seconds is near enough.

How does it work? In the top of the funnel there is a mesh and just below that a ring around the funnel. With your finger over the hole in the bottom of the funnel, fill it with mud up to the ring via the mesh, then get a helper to clean out the cup. With a stop watch at zero, take your finger off the hole in the bottom and fill the cup to the ring which marks off one U.S. quart — you've got your viscosity reading.

Do this at the start of each shift or when you have mixed fresh mud (remember, if your mud is bentonite it should not be used for at least twelve hours) then note it in your daily log.

Every hour thereafter measure the viscosity again at the point of entry into the suction hose, noting each reading in the log. If it starts to go down then you probably have dilution by groundwater (sand can have a similar effect) but if it starts to go up then check for two things.

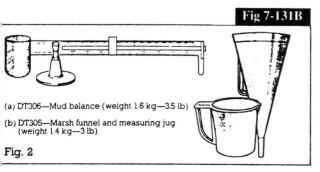

Fig 7-131B

(a) DT306—Mud balance (weight 1.6 kg—3.5 lb)

(b) DT305—Marsh funnel and measuring jug (weight 1.4 kg—3 lb)

Fig. 2

1) There could be clay in the hole and you are adding it to your mud. This is not a good thing because if it gets out of hand (too thick) you will have trouble dropping out cuttings and your pump and swivel etc. will suffer. Mud thinning agents are available in most countries — take advice.

2) The mud is carrying solids (cuttings). This is really serious and the size of pits, channels and stilling pools (see figure 4-84A) should be looked at immediately and adjusted as required.

How can you confirm that the mud is carrying solids? Well, you need an instrument which is also essential to every drilling rig; it is a:-

Mud balance (figure 7-131B)

What does it do? It weighs the mud, giving you weight in pounds per gallon (ppg), weight in pounds per cubic foot (ppcf), specific gravity (sg) and pressure per thousand feet (psi m/ft).

The measurement we are most interested in is the specific gravity which, of water, is 1.0. Therefore your mud should be reading something above 1.0; read it carefully.

How do you use it? Put the mud balance on a level surface, fill the cup with mud, giving it a little tap to get rid of bubbles, put the lid on the cup, wipe away any excess mud and move the cursor along until the bubble (spirit level) shows that the balance is level. You then read off the sg (or whatever you like) from the side of the cursor, which is marked with an arrow.

This is done at the start of each shift or when making fresh mud, (weighing the mud at the point of exit from the hole and at the point of entry into the suction hose) and every hour through the shift, noting the readings in the daily log.

You will now see that if you have measured and weighed your mud when fresh, all subsequent readings can be compared with that figure and you can judge what it is doing and adjust when necessary; the reading at the start of each shift is also compared with the "fresh figure" for the same reasons.

So, if the marsh funnel reading at the suction is showing higher than normal and you don't think there is any clay in the hole, then the mud balance will also show a higher reading. Therefore you are carrying cuttings through with the mud. If you are, and your mud pits are correct, then there is only one course of action: *dump the mud* and mix a new batch.

Why do you have to take measurements with the mud balance at the points of exit and entry? Simple, to make a comparison, ensuring that you are dropping out cuttings and that your pump, swivel etc., are protected.

Now we have a measure of mud control and, to summarise, if the viscosity goes down you replenish the mud to get it back to what you want, but don't forget that you can't just add bentonite and use it straight away — you can with polymer muds. If the viscosity goes up, you check for added clay or solids-carrying with the mud balance.

Make sure you enter your readings in the log, because it is no good taking readings and having nothing to compare them with. Correlating mud viscosity and weights (sg) is fascinating and will give you a lot of information.

Hydrocyclone

This is an expensive piece of equipment which, if you can afford it, makes for good mud control and will often allow you to use smaller (not much) mud pits and tanks.

It is like a giant de-sander (figure 7-131A) which must have its own pump, a centrifugal? It works similarly to the de-sander but on a much bigger scale.

Have a word with your friendly supplier.

Fishing tools

A kit of fishing tools with the rig is essential and they must be carried at all times. There is an old saying "If you've got 'em you won't need 'em, if you haven't you will".

If it were possible to list every different type of fishing tool ever made you would fill a book a thousand pages thick, at least. Old hands at the game will know what we are talking about so, because of this, we are restricting this section to listing the three basic tools you should have in your armoury and then giving a few case histories of fishing jobs done to illustrate what can be done when required.

1) *The fishing tap*

Sometimes known as a carrot or spear, this is a male tapered tool (figure 7-133A) which has a fine right hand (left hand for left hand string) thread machined onto it and the whole thing hardened above the hardness of tool joints (etc.) so that it can bite into already hard metals. It must always have a hole through it, because you must have the option to flush when fishing; it is finished off with a thread at the top to match your drill-pipe.

The tap can be stepped to enable its use in different diameters (figure 7-133A) or you can have separate fishing taps to accommodate other items in your drill string. Some manufacturers have special fishing tools for their products and these items should always be considered.

The idea behind this tool is that, once the "fish" is known and it is decided that this is the required tool, it is attached to your drill-pipe and entered into the "fish". Weight and rotation are applied and the tap will cut its way into the "fish" which is then caught and pulled out of the hole. Easy? — See later.

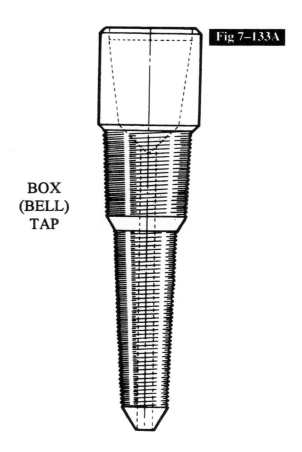

Fig 7–133A

BOX
(BELL)
TAP

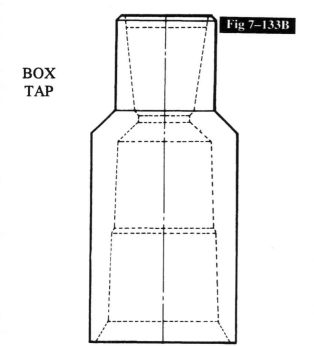

Fig 7–133B

BOX
TAP

2) *Bell tap*

This is the female version of the fishing tap and is threaded on the inside, and if this is the tool you need then you use it the same way except that it goes over the fish and not into it. (figure 7-133B).

It is also known as an "overshot".

3) *Magnet*

This can be either solid state or an electro-magnet. The latter is, or should be, simple in design and made to work off your truck or rig battery.

If you don't have a sand-line winch (or wire-line winch) on the rig, a strong hand-winch can be used to run the magnet into the hole. After all, you aren't going to pick up a string of drill collars with a magnet, are you! Make sure you have wire-line on the winch and not manila rope as we saw used once — we saw it because we had to fish it out, broken bit, magnet and rope; the rope had parted.

The diameter of the magnet should be judged according to the bits you are using, and don't forget that there are impurities in most forms of tungsten carbide therefore it is magnetic — try it and see. Broken tungsten carbide left in the hole will break new inserts — get it out.

The winch is most important, otherwise making several "trips" with a magnet on the end of drill-pipe will only add to your frustration.

With the electro-magnet, the electric wires will, of course, have to be of sufficient length for your hole programme and be *watertight*.

Fishing

So what is fishing? It is getting broken things out of the hole — the broken thing is called the "fish".

It is ten percent total calm and ninety percent third degree hysteria.

It is heart stopping, gut jerking, frustrating to the "nth" degree. It is depressing, stimulating, it breaks up friendships (even marriages) but a successful fishing job is thrilling to the point of orgasm.

What do you do then when you have a problem? This:-

1) You sit down calmly and make yourself a cup of tea. This not only tends to calm you down but also gives the crew confidence in you.

2) Go back to the machine and lift the remaining tools off the break, noting the weight of the tools in the hole on your weight indicator *and writing down that weight*; you might forget it in the frustration that is to follow.

3) Trip the tools, noting the depth in your book, and examine the failure.

4) Decide on the type of fishing tool to be used and run it into the hole on drill-pipe (winch if magnet). If it is, say, a drill-pipe "twist-off" (breakage), locate the fishing tool into, or over (tap or overshot) the fish, give it a twist and a push then pull back slightly. Now check your tool weight again on the weight indicator and if it is greater than the weight you have written down up to your estimated total tool weight, then the "fish is hooked" (if not, try again).

5) Now carefully trip the string of tools, noting your tool weight as you go; any sudden drop in weight and you might have lost the fish and you

will have to start all over again — don't forget the tea.

6) If necessary flush to help ease the fish (remember the hole through the fishing tool?).

7) If your "fish" is broken carbides or the like, secure the magnet to the winch line (remember, *secure*) and make a pass, matching any particles brought up to the broken bit until you think you have it all — broken carbides left in the hole will break up a newly introduced bit.

You might have to make further passes with the magnet to get the hole clean and even, at times, to run your tools down to wash around the "fish".

Junking

We ought to mention this briefly because, whilst it is mainly used in the "oilfield" for deep holes where it would be unthinkable to use a magnet, some of you might be involved in deep water wells.

Junk? This can be bits of tungsten carbide, or a broken drag bit or, indeed, any other sort of debris, even a hand tool.

Junk mill? This is a sort of a bit with multiple rows of tungsten carbide teeth throughout the diameter. This tool is fixed on the end of your drill-pipe, lowered onto the "junk", rotated and pushed until the "junk" is ground up and flushed to surface — there is a myriad of types of junk mill.

A junk basket? Is a bit like a tube with holes in and open at the top and is fitted immediately above the junk bit (sometimes even, run as standard) so that those larger pieces of junk that are too heavy to run to surface (until ground up) in the mud (etc.) might just hover above the bit and very conveniently drop into the junk basket and come up on the next trip.

Case histories

We have already touched on one, and that is the case of the broken manila rope.

The rig was located 10,000 feet down a mine drilling angled holes 300 feet deep from one level to another; a simple job because *it was hard rock* and a down-the-hole hammer was being used.

Within 10 feet or so of TD the shank broke off the bit, leaving the rest in the hole after it was decided to trip the tools to see why there was no penetration.

The driller ran the magnet down on the end of the manila rope, got a hold on the fish, gave it a tug and the rope parted — we were called in. We had a broken bit, a magnet and the manila rope to fish.

We got a piece of sturdy wire and put a turn in it, leaving a gap in the turn just big enough to accept the diameter of the rope; the wire was brazed on to and old sub. and the whole thing run into the hole on drill-pipe, the theory being that we would stab the rope and hook it into the turn.

We gave the string a good push into the pile of twisted rope in the hole and tripped out. Not only did we get the rope, we got the magnet and bit as well — first time.

Drill collars aren't too difficult by the way, as, if used correctly, their diameter is not too much less than the bit. Therefore, if they twist off, the hole in the middle is near centre of the drilled hole and the location of the fishing tool (male usually) is relatively simple. But, if you are drilling large (relatively) diameter holes on small drill-pipe, that is where trouble starts, if the pipe twists off. This happened to us:-

The location was West Africa and we were drilling a 650' 9⅝" hole with 6" drill collars and 2⅜" externally upset drill-pipe.

The drill-pipe had a history. It was second hand discarded pipe which was badly bent and had been cut down into short lengths and the tool joints welded on with just a straight weld.

Anyway, the pipe twisted off at 460 feet at the weld under the box end tool joint (we were working pin down), leaving a twisted bit of 2⅜" pipe in a 9⅝" hole — and we didn't have an overshot nor any material or facilities from which to make one, we did have some 6" casing and a welder; the nearest workshop was a day's drive away each way.

Inside the bottom of the somewhat flimsy casing we welded some bits of old (thin — that's all we had remember) sheet steel in the shape of a taper, the ID of which was just bigger than the OD of the tool joints: we ran this into the hole attached to the 6" casing.

As we feared, the plain end of the drill-pipe was lying on the side of the hole and in such a position would have been impossible to locate with our fishing tap; we felt it enter the casing. We slowly continued the downward movement of the casing until we felt the next tool joint (externally upset, remember?) enter the taper in the casing "shoe" and there we stopped and clamped the casing, our theory being that with the tool joint in the shoe the plain end would be somewhere near the centre of the casing.

With the fishing tap attached, we ran the drill-pipe into the hole and almost immediately located the broken pipe; we turned and pushed. We "weighed" the string and the weight indicator showed that we had hooked the fish.

The drill-pipe was disconnected from the rig and the string left to sit on bottom. This is not good practice and should never be done except in such dire circumstances as these.

The casing was withdrawn, the drill-pipe was attached to the rotary head and we slowly, very slowly tripped the tools. Not a word was spoken.

Constant monitoring of the weight indicator showed we still had the "fish" and hearts leapt into mouths every time there was even a slight "snag" in the hole.

We all watched fascinated as the fishing tool, complete with fish, passed the working table. We gently clamped the string to the rig below the fishing tool, cheered, and had another cup of tea.

There is a moral to this story. If the owner of the equipment had had an "overshot" (bell tap?) with the rig it would have saved us a lot of trouble — he got one later.

The last of this trilogy of case histories concerns a core drilling job in the Middle East, where we were experimenting with taking core samples of soft, friable phosphates (the outcome was 100% but that is another story and another book); we were working with a very experienced French crew.

The driller had just tripped the full core barrel from the best part of 1000 feet and the barrel was held in the working table whilst the last joint of drill-pipe was being racked. Unfortunately, the device holding the barrel in the table was of very poor quality and the core barrel fell into the hole reaching bottom much faster than it came up.

The driller's expression did not change, he

didn't even look round, he just ran the pipe back into the hole (there was no breakage, therefore no fishing tool) touched the fish, made a mark on the drill-pipe with chalk, measured the length of a pin thread and made another mark on the drill-pipe, the difference between the two chalk marks being the length of the thread.

It is important to remember here that the core barrel, even when full, would hardly move the needle on the weight indicator, therefore he had to have another reference.

He relocated the drill-pipe and slowly screwed in watching his torque gauge. When the second chalk mark indicated he had made up the thread, the torque gauge gave a very slight "kick" — he knew all was well, tripped out, never smiled nor said a word about it. He later modified the working table. By the way, Frenchmen don't often drink tea.

There is a moral to this story as well. The difference between the diameters of the core bit and the core barrel is just a few millimetres, therefore his chances of hooking were good from the start. This is similar with drill collars and with down-the-hole hammers.

To summarise fishing. Keep calm, think long and deeply, take all the help and advice you can get, and if you don't catch the "fish" first time, try again. But, and this is a big but, you have to know the value of the hole drilled so far against the cost of the time taken to fish and decide if it is preferable to start another hole.

Most manufacturers of special tools like down-the-hole hammers will have a kit of fishing tools available — get them.

Break-out tools

These have been touched on already, but we thought that a little more expansion would be wise.

They should be called break-out/make-up tools, because drill collars, drill-pipe, subs. and just about everything else that goes in the hole must be tightened and the supplier of the tools must give you the figures of make-up torque; to think that tools will tighten themselves in the hole is a grave error.

If you ever see pin end tool joints that are cracked or broken off it is almost certain that the crew are not doing their job correctly. *They must be tight.*

And don't forget — every action has a reaction, so don't think you can get an enormous tong around a tight pipe and expect a flimsy laykey (slip plate) to resist the pull.

So what are the options? Power tongs, where a pair of "wrap around" tongs operate under their own power; manual tongs (figure 7-137B) which are most desirable; stilson (pipe) wrenches, a cheap substitute for tongs: laykeys (slip plates — see figure 7-137A) which are excellent when drilling very shallow holes; hydraulic clamps, not our favourite as already said previously; or half a dozen people on a pipe.

We can discount the first because of their enormous expense for a water well operation, the last for being a little bit frivolous, although still not unknown, and hydraulic clamps because of their inherent danger in being somewhat unreliable in the event of a hydraulic failure. We like things to have a "positive" action because they are usually safer.

Let us stop beating about the bush here and give you our recommendations, which relate to the depth being drilled.

If the depth drilled is less than, say, 300 feet and your drill-pipe has "flats" machined in the tool joints, (to take a spanner without weakening them or restricting the size of the hole in the tool joints), then a strong laykey (slip plate) held firmly in the working table (figure 7-138A) and a manual tong counterbalanced over the mast is a good and reliable set up for a top drive rig. This can be adapted for a rotary table.

Deeper than that, and automatically for rotary table rigs, you must consider slips to hold the pipe and a pair of tongs to make up and break out: stilson (pipe) wrenches can be substitutes for the tongs but are not as effective due to their "two point" loading on the pipe, which can cause damage.

We are, of course, assuming that your rig has a break-out arrangement, such as a hydraulic ram or a cathead or something. If it doesn't, then the idea of half a dozen people on the end of a pipe is not so frivolous.

Let us think about tongs for a moment. They "wrap around" the pipe, (or drill collar — you've got to have them), with a number of replaceable jaws (number according to size and manufacturer) and give a multi-point loading on the pipe etc. They will have a range of pipe/collar sizes to which they can be applied by altering the jaw sizes; does that make sense? If not see figure 7-137B.

The handle (?) of the tong will have a location for the attachment of your break-out system and it will be towards the outside. Should we remind you of the laws of force, in that the further you put the "pulling" device out along the handle the greater will be the force applied to the object?

We have too often seen crews using a stilson (pipe) wrench with the pulling cable next to the jaws and the object not moving. A simple shackle, or "D" clamp, of sufficient capacity fixed at the outside of the wrench handle with the wire from the break-out tool attached thereto would (and usually does) "break-out".

OK, so tongs can be expensive, but they are well worth it. They can be a little on the heavy side, that is why they are counterbalanced with a weight over a sheave. On the point of weight, investigate oil-field aluminium tubing tongs — but don't forget sets of jaws.

Fig 7-137A

Fig 7-137B

Fig 7-138A

Rig lighting

Have you ever tried looking up a mast to make-up drill-pipe or see your kelly run back, against those ghastly lights some manufacturers fit into the crown block (mast head) — you can't see a thing. Lights should be focused on objects not on the crew.

Lights to illuminate the working area around the rig should be remote from the rig and those mounted on the rig mast should illuminate the working/rotary table, the driller's controls, the rotary head/kelly and crown block, and should be *shining* away from the crew's eyes.

* * *

The next part of this book will be considered by some to be somewhat academic for the water well industry describing, as it does, blow-out preventers. But don't be fooled; we have known crews lost and equipment destroyed because drilling companies thought that and went into known oil and gas areas (etc.) unprepared.

We have seen massive artesian flows from water wells that were almost uncontrollable, when the contractor knew the area was one of artesian conditions.

Just give this next section a chance. Read it, then give it a lot of thought.

Blow-out preventers

Rule No. 1 — the blow-out preventer is only as good as the cement job you've done on the casing, because it (or they — see later) sit on top of your casing and if you have to close-off because of pressure, and the cement isn't any good, then one rig goes straight into orbit.

Cementing was the subject of an earlier book. Therefore, for this section, we will assume that you've done a good cementing job.

What do the b.o.p.'s (blow-out preventer) do? They are there to help control abnormal formation pressures.

Doesn't sound very exciting does it? You thought we were going to say that they stop oil shooting hundreds of feet into the air. Very dramatic — well, they do that too because that's what an abnormal formation pressure is; i.e. a situation where the formations produce a higher pressure than the normal for that particular depth and condition.

This applies to water as well as oil and gas. We have seen, on numerous occasions, a spout of water gushing out of the hole to more than thirty feet in the air and the crew helpless to do much, *in areas where artesian pressure is known to exist.*

We are then called in to control the flow, usually with heavy muds, but in these circumstances it is costly, because a great deal of time is lost whilst drilling engineers arrive. Don't forget that engineers have to be called from somewhere else — time, in drilling, is very expensive. See Book Nine for costing formulae.

Now let us do that job all over again. This time we have a nice little b.o.p. sitting over the hole and suddenly we find more mud coming out of the hole than there should be (the first indication of abnormal formation pressures). The driller signals the well trained crew; the first of whom unscrews the "diverter valve" (see later) while the second closes the b.o.p.

Wonderful. The artesian water is now flowing out of the blooey line (see later) well away from the rig, and preparations are made for the "control". Isn't that nice? Additionally, one can effect a "balancing act" with in-hole pressures through "chokes" (see later) which can also be used to "kill" the hole with cement if problems are too great — all nice and clean and sensible?

Not quite. You see — if formation pressure is so high that you should be running a b.o.p. Then it is high enough to run up the drill-pipe and back through your pump as well. Therefore we must do something about that as well — this something is called a drill-pipe safety valve.

This is a miniature b.o.p. which fits permanently under the rotary head or on the kelly. It is in effect a rather superior non-return valve, allowing the normal passage of drilling fluids (downwards) but preventing backflow.

B.o.p's are not as expensive as you might think — in fact they are usually rented, and any client worth their salt will accept arguments, therefore expenditure, in favour of such equipment in the right circumstances.

There are three types of b.o.p. in general use. They are:-

1. *Ram type*
For our purposes, this is the most commonly used. A pair of hydraulic rams, manually or power (or both) operated, are contained in a casing (figure 7-140A). The rams will be designed to fit around your drill-pipe when closed, thus shutting off the well. The rams can also be "blind", meaning blank. Thus, when there are no tools in the hole, for instance, these rams can be closed, thereby shutting the hole off. This is a "single drilled" unit, API nomenclature "R".

A casing can contain two pairs of rams — say, one for pipe and the other blind. This is known as double drilled, API "Rd". In such circumstances the blind rams would be on top. If you have two single drilled b.o.p's together, the top would have the blind rams. Why? Because, in the event of trouble, you can close around the pipe and work on the units above that — if the bottom rams were blind you couldn't do that.

You must always have a minimum of two sets of rams, pipe and blind, for obvious reasons.

B.o.p's are specified in internal diameter and pressure. If, for instance, the b.o.p. is 13″ then the internal bore through the b.o.p. unit will be 13″, thus allowing the passage of tools up to that diameter.

The pressure rating for the b.o.p. unit as a whole will be, say, 2000psi, 5000psi, 15000psi, etc. API nomenclature 2m, 5m, 15m, so on and so on.

You must specify your own ram dimension according to drill-pipe dia. etc. You can have several b.o.p's, in a "stack" (that is the correct word to use — b.o.p. stack), all single drilled, all double, or a combination of the two. They will include "pipe" and "blind" with alternatives for

Fig 7-140A

BOP STACK — EXAMPLE

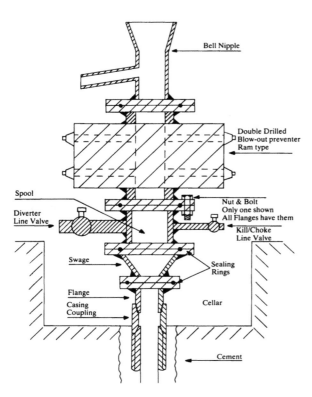

Bell Nipple

Double Drilled
Blow-out preventer
Ram type

Spool

Diverter
Line Valve

Nut & Bolt
Only one shown
All Flanges have them

Kill/Choke
Line Valve

Swage

Sealing
Rings

Flange

Casing
Coupling

Cellar

Cement

(Not to Scale)

drill collars, kelly (see later), and even one that incorporates cutters that will cut off the drill-pipe. In water wells though, lets be sensible and look at simplicity because going overboard would present an unwieldy stack demanding enormous clearance under the table.

2. *Annular type*

Always hydraulically operated, this is a sort of squeezable rubbery material that closes around the tool inside it and is correct to use for inhole tools of an irregular shape; it will usually sit on top of the "stack". It is signified as "A" by API.

For the control of fairly low pressure artesian water a number of small manufacturers have made this type of b.o.p. in a quite simple form. They create a pressure chamber with a hole through it large enough to take the required tool sizes and with flanges top and bottom. The hole is closed off by a rubberised (neoprene or similar) sleeve bonded top and bottom, through which the tools pass. You have an inlet from the by-pass system on the mud pump and an outlet back in to pits; inlet and outlet are valved.

If your mud pump is big enough, you can have mud by-passing the b.o.p. all the time and, in the event of trouble, close the outlet which will pressurise the system, inflate the diaphragm, and close off the hole. Please remember to shut the inlet — bypassing the mud at the pump — once inflation has been achieved. This reduces the system pressure and so prevents the pump relief from blowing.

With small pumps you will have to be quick and pass mud into the chamber by opening the inlet valve in order to get the "shut off" of the offending pressure.

You must always have a diverter under this b.o.p. In fact it is recommended that a diverter should be used under *all* stacks.

3. *Rotating head*

There are lots of colloquial names for this unit, so let us describe it and leave you at liberty to call it what you want; you more than likely won't need it anyway so it is pretty academic.

Again — it is a rubberised unit that is always closed around the drill-pipe/kelly (but in this case rotates with the pipe/kelly) and it is used mainly for air/gas drilling. Under the rotating head is a by-pass to take away air or gas mixed with cuttings. Very skilled operators will use this with "underbalanced" muds, allowing the hole to do the work.

The tool is a bit like a swivel, because there are packings between the rotating part and the stationery.

In API jargon this is named "G".

And now for other items which go to make up the stack. Please refer to figure 7-140A as we go along, starting from the bottom and working upwards.

Flanges

We use flanges to connect the various items in the stack. The vision of tons of iron (b.o.p. etc.) being screwed together is not an appealing one. Flanges must be of equal pressure rating to the rest of the system and made to API standards.

The one bit of "screwing" we like to see in the stack is the first (bottom) flange which screws onto the casing — weld-on units are also available. We prefer to use this rather than a conventional casing head (see figure 7-140A) because, being the "poor" relation in the drilling industry, we might only have one b.o.p. and various sizes of casing. So what do we do?

Well, a casing head will have two outlets, one for the diverter (blooey) and the other for the choke system (figure 7-140A). This item is expensive and it fixes directly to the casing. If you have different casing sizes to consider then using a swage on the casing flange followed by a spool having the same overall dimensions as the b.o.p. is the most economical solution.

The mating faces of all the flanges should be machine grooved to API standards to accept ring seals.

Swage

A swage is a unit of two dissimilar flanges connected by a piece of pipe and made to to API standards. It must be similar, or better, in pressure rating to the b.o.p. in other words it is a very sophisticated adaptor, one flange to suit the casing flange and the other to suit the spool.

So, where are we now? Oh yes, casing flange, ring seal, swage (if required) then ring seal then:-

Spool

What is this? It is a pair of flanges separated by a short piece of pipe, and from the pipe are usually two outlets, both with valves. One pipe goes to your diverter line (blooey) and the other to the choke system. The whole is unit built and tested to pressures commensurate with the pressure rating of the b.o.p. or more, including the valves.

The flanges, therefore the bore, must suit the flanges on the b.o.p. and again, be made to the exacting API standards.

In API language a spool is "S".

The blow-out preventer

This can be manually or hydraulically operated or both. If you are dealing with a gas zone, then hydraulics, with a remote control are essential, but if the control is for water then manual operation is ok if you don't mind getting a bit wet; the longer the operating arms, within reason, the drier you will be.

As you need blind and pipe rams (at least) why not have one double drilled b.o.p. eh? It makes it a bit easier to handle overall.

So, you've put a ring seal between the spool and the b.o.p. and tightened these down with the correct sizes of nuts and bolts (the tightness of these nuts and bolts must be checked constantly whilst drilling etc.) what next? Another ring seal.

Now, we can't have mud coming up the hole, through the b.o.p. then spilling over into the cellar, can we? So we next have a "bell-nipple".

Bell-nipple

Figure 7-140A shows one of these and you can see how simple it is. A flange, with seal, connects it to the b.o.p. (remember we are putting a "simple" stack together).

A bell-nipple comprises a vertical pipe, "flared" at the top to ease entry of your tools and to slow down excessive flow of fluid. Out of this comes another pipe, usually at a slight downward slant, and this is your flow-line which takes your mud into the shale shaker or pit channels.

Remote b.o.p. control

Here a word of warning. If you are working in an oil and gas zone, check the prevailing wind direction and set up everything, including your camp, so that the wind will take any gases etc. away from the rig, camp and remote b.o.p. control.

A typical b.o.p. remote hook-up would comprise a couple of valves, a hydraulic power pack and a couple of pressure cylinders known as accumulators (one for each pair of rams) plus connecting hoses into the b.o.p.

As our b.o.p. is a double drilled unit, we have to have a separate operating valve (open/close) for each pair of rams. Now, if the operation is purely hydraulic, the time taken for the rams to close is quite long and, if the hole is troublesome, can be too long, so the hydraulics are "boosted" by nitrogen. The accumulators contain nitrogen and hydraulic oil separated by a float.

The hydraulic pump, which is an integral part of the unit and must be obtained as a whole, will pump hydraulic oil into the accumulators up to a pre-determined pressure (then the pump cuts out) thus compressing the nitrogen. We are sure you can now guess what happens when you operate a valve — those rams shut like the wind.

The b.o.p. must be tested daily and, once tested satisfactorily, the hydraulics *must be pressurised again*.

Keep the remote control panel as far away from the rig as possible.

Diverter line (blooey)

This comes out of the larger of the two valves in the spool, the diameter of which is determined by the diameter of the spool, which is controlled by the diameter of the b.o.p. etc. etc. — but 4"–6" looks good — the bigger the better. The valve is, of course, high pressure but the blooey needn't be — good quality seamless steel pipe is adequate.

Don't forget — lay out in line with the wind and make it about 100 feet long. Anchor it down, because a bit of pressure in it will make it "take-off".

So what does "old blooey" do? Imagine you have just struck high pressure water in the hole and you have shut the b.o.p. — what do you do then, just sit there? No, you open the "blooey" (diverter) and out goes the water (for actual procedure see later). Sometimes the pressure will go down and all is well. You then open the b.o.p., recharge the hydraulics, and carry on.

If the pressure doesn't reduce, then you have time to think about what to do next (see elsewhere for heavy muds etc.), without adding to the problems in the hole. Mind you, if you are a very superior type of driller you might want to use that pressure (a sort of balancing act) so you would go to the other side of the spool to the second, and smaller, high pressure valve where you will have your:

Chokes

Chokes can be manual or automatic (remote) and what happens? You shut the blooey and adjust the choke system until the back pressure in the hole is released back into the mud pits/tanks in a controlled fashion. This way an "overbalancing" of the inhole pressure may allow drilling to continue.

As a part of this system, or indeed, as a replacement for it, you can have a "kill" line. All that means is that, should you have a "difficult" well, you can connect in a high pressure pumping system and "kill" by introducing the relevant fluids.

Some b.o.p. units have an integral connection for the chokes, meaning that you can use the spool outlet for "the kill" and that connection for the chokes. Or you can have "the kill" opposite the chokes in the spool — the choice is yours, and there are plenty of them, but safety is paramount.

Crew training

The crew must be aware of the exact procedure to be adopted in the event of trouble and they must be rehearsed daily, sometimes in a mock emergency. Each person must have their function, with a back-up in the event of absence. Our own favoured procedure is:-

1) Driller gives the order to close off.

2) Crew-hand opens diverter valve (blooey).

3) Another crew-hand closes appropriate rams *simultaneously* with opening of diverter.

4) Crew abandon rig to a safe distance to consider situation, running into the wind — discharge from the blooey could be dangerous.

This whole operation should take only seconds — rehearse!

Be warned: in an oil and gas area you will be required to have additional safety equipment such as a gas "sniffer", gas detector and personal safety equipment (air packs) etc., and such decisions are made by discussions between you and your client; do not short-cut here.

The cellar

To fit a b.o.p. stack under a rig, you either have to lift the rig up high on a ramp, dig a hole in the ground to accommodate it, or use a combination of the two. The hole in the ground is called a "cellar".

The depth of the cellar is determined by the height of the stack, plus the height the bit of casing sticking out of the ground, plus a reasonable distance between the top of the bell-nipple and the bottom of the table. When excavated it should be nicely cemented and its lateral dimensions should allow for the width and depth of the b.o.p. plus a bit to allow movement around.

With a water well rig, it is likely that the dimensions of the cellar could weaken the ground under it — be careful, you don't want a rig in your cellar do you?

There is a definite procedure we like to adopt for starting a well where a b.o.p. is required, it is:-

1) Set up your rig and drill to casing depth.

2) Case off, measuring each length of casing accurately and carefully marking your landing joint so that you can pin-point the position in the ground at which the casing thread — on which the stack is to sit — will be positioned.

3) Clamp the landing joint in the rig securely and cement off. Do not use slips in this instance because, after cementing, you will have no upward movement of the casing to get them out.

4) When you are satisfied that the cement is dry, excavate the cellar, exposing the casing thread for the stack, and remove the landing joint carefully.

5) Cement the cellar.

6) Install the stack, making sure that all nuts and bolts are tight. Refer to figure   for the procedure for our "simple" b.o.p. stack and if there is any variation from this substitute the required items, making sure that you include all sealing rings and that pressure ratings and API standards have been satisfied.

Assuming that our simple stack is rated at say, 2000psi and is 13″ ID you have a "2m — 13″ — SRd" stack, remember? 2000psi rating, 13″ ID, and spool/double drilled b.o.p. — simple eh?

Always keep the cellar clean! If you have to put "duckboards" over it remember that *all* valves must be exposed for immediate use in an emergency. If you are using manual closing of the b.o.p. make sure that the "knuckle" joints are well greased and that rods and handles are not obstructed in any way. *Lives could be at stake.* Test every day and, if using hydraulics, make sure pressure is up to specification and test the nitrogen; there will be a little valve on top of the accumulator for this purpose.

Now run your tools in the hole and... no, you are not going to drill yet.

With the bit near bottom, close the pipe rams of the b.o.p. around the drill-pipe and turn on your mud pump. Get the hole/b.o.p. up to pressure and check for leaks and, indeed, any casing movement.

If all is well, open the pipe rams and re-pressurise the b.o.p. system. If all is not well effect repairs — and now... drill on.

There are a few other auxiliaries you will need, such as a water level indicator, sample trays, shovels, a mirror to look down the hole using the sun etc., but they will come either with experience or on the instruction of the geologist etc. so we will not dwell on them at this stage.

However, there are still a couple of items still to be explained and they are:-

Foam tanks

It is not uncommon for people to use two drums with the tops cut off for use as foam tanks; one drilling whilst the other is being prepared. This is cheap and convenient but, for a large scale operation there is a better way.

Use a tank of large capacity (we favour one metre cube) which has a divider to make it into two separate tanks. Into each tank, at the bottom, you put an outlet tap, and the taps are joined together by a piece of pipe open at one end only. The suction hose from the foam pump is connected to this opening or, indeed, the foam pump can be mounted on a platform at this point and direct-coupled to the pipe.

Fill and mix both tanks. Open one valve and start drilling. When that tank is empty close the valve and, at the same time, open the other; you are now using the second tank whilst mixing the first. It is quite normal to mount a mixer on top of the tanks which can act as an agitator as well.

Figure 7-144A illustrates this type of tank unit.

Deviation checking disc

No hole is straight but it must be straight enough to allow for the correct installation and operation of in-hole equipment. Strict limits are imposed in many countries. For instance, in the United Kingdom the maximum allowable deviation is four inches in one hundred feet. Over that you redrill.

Deviation must not be confused with deflection. Deflection is a controlled situation and deviation comes about, mainly, by poor drilling practices.

To check out the hole when it is finished, there is a very simple method which is as follows and uses a winch (usually hand operated) with sheave and "dolly", a disc marked out in degrees and a set of tables. Let us take them separately:-

The winch, sheave and dolly

The winch should be equipped with a thin wire-line and a brake. A sheave is suspended under the rotary head (or on the hook) so as to enable the

Fig 7-144A

wire to run into the centre of the hole. Hanging on the wire is a "dolly" slightly less in diameter to the part of the hole to be surveyed — it might be the casing or the hole itself, both of which have different diameters.

The centre of the sheave should be ten feet above the working (rotary) table with the wire-line running exactly in the centre of the hole.

The deviation disc (see figure   )

From figure 7-145A you will see that the disc is accurately measured out in concentric circles, each with a value in degrees. A slot is cut from the centre to the circumference, wide enough to take the wire-line. The wire-line must be marked off in feet.

144

With the "dolly" suspended just inside the well, the disc is placed in the table and mounted so that is free to rotate — for reasons that will be explained. The wire rope runs into the slot cut in the disc and must be at the centre of the disc, therefore at the centre of the well.

Fig 7-145A

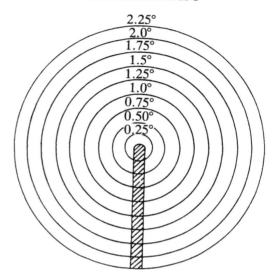

DEVIATION DISC

Deviation tables

The "dolly" is slowly lowered into the well, and when there is deviation, the wire will move towards the direction of deviation and the disc should move or be moved so that the wire-line is not encumbered in the slot. With the winch brake applied, read off the value in degrees from the concentric circles against which the wire-line has come to rest, and the depth in feet off the wire-line. Now compare those readings with the deviation tables in figure 7-145B.

If the wire-line rests between the concentric circles then extrapolation is permissible.

An example of a reading is shown in figure 7-145B.

By referring the direction of deviation to a compass, orientation is possible and the well can be plotted.

Fig 7-145B

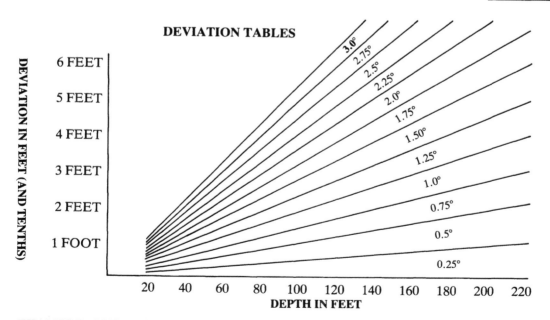

DEVIATION TABLES

EXAMPLE (OUTLINED):– 2 Degrees of Deviation at 86 Feet is equal to 3 feet of Deviation

Extrapolation Permissible

GENERAL VIEW OF ARRANGEMENT FOR DEVIATION CHECKING Fig 7–146A

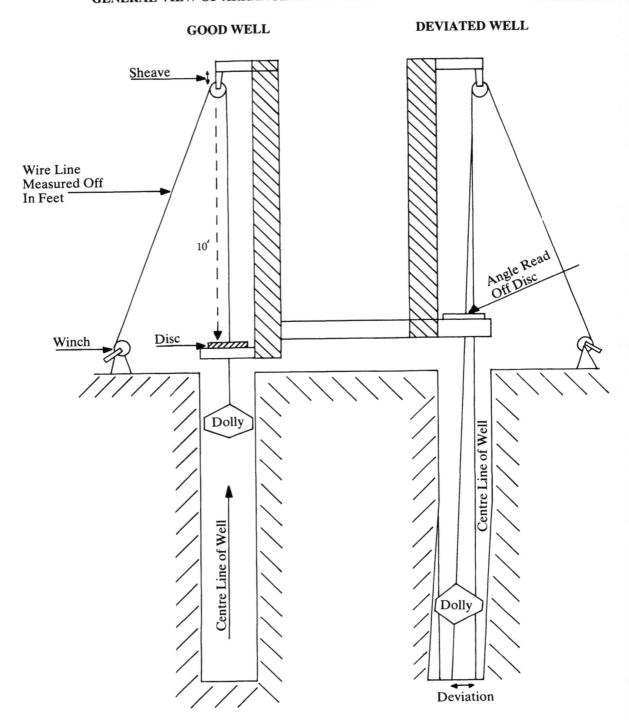

8 Basic Geology

Foreword to Book Eight

We are not geologists so what follows is a list of some of the geological formations in which we have worked, the tools we have come to use in them and hazards we have found therein.

We suppose this, being the last book in the series concerned with field functions is correctly placed because it acts as a sort of summary of all that has gone before.

One thing we hope it will do is to help you to pre-plan your operation, thereby getting good overall performance.

Good drilling!

Basic Geology

Sometimes you will find a geologist who is a bit pompous and talks down to a mere driller like you, giving only text book names to a bit of rock. We have therefore gone into a little bit of detail when it comes to naming formations and used a little bit of cross referencing.

To reiterate for a moment though, "the harder the drilling the less difficult it is, the softer the more difficult". That must be the second most important "rule" in the world of drilling, the first being "if you can't clean the hole don't start". Have those two painted on your rig in full view of all.

Mother nature is unpredictable and will come up with surprises at the wrong moment, just when you thought you were doing well. It is rare to find gifts from her which are appropriate to a lazy mind. Mind you, she has given us some areas which are "easy" but they are rare. Let us tell you a story which will illustrate these points.

One of our engineers was working in a rather difficult area of India with lots of "overburden" which had to be cased off, followed by hard rock; it was taking about two shifts to complete a well.

An official came to him and said that his production did not compare with a similar rig working further north; in fact the other rig was doing twice the work. Thinking that he might learn something new he carried out an investigation.

The area being drilled by the other rig was pure rock, with only a foot or two of "overburden" which could be ignored; it was like drilling a quarry. His down-the-hole hammer drilled it without interruption at great speed. Also he could complete most of his wells at minimum depth. In fact his production was very poor for the conditions.

Is there a moral to this story? Yes there is. Don't be frightened to note drilling, and other problems on your daily reports, otherwise people can get the wrong ideas. The more "head-office" knows, the better chance you have in getting better equipment.

Now let's get on with the business of geology. There are three main classifications of formations, they are igneous, sedimentary and metamorphic; let us look at them separately, with notes.

Igneous

These rocks have been created from a molten mass and can be classified into those that cooled slowly deep into the earth and those which underwent a quicker solidification nearer the surface. The former will be coarse-grained — such as granite, diorite, gabbro — and the latter fine grained — such as basalt, andesite and rhyolite.

In general they are on the "easier" end of the drilling spectrum, being harder and dense, but they can be full of surprises, as we will see. With one or two exceptions, water is found in cracks, and the more cracks the better, but don't expect there to be water in abundance in themselves.

By great and careful work, we have made igneous rocks yield sufficient water for a modest pump, but generally don't expect to produce much above hand-pump discharges. Even then, the actual location of a well will be a bit hit and miss, but don't lose heart. Read on.

Granite will be coloured light grey to reddish and the colour will deepen to dark grey for the diorite and gabbro. Colouring for basalt is very dark green/black and this rock will have a higher specific gravity than the courser grained granite.

Whilst igneous rock masses will tend to be hard, there is a degree of weathering. In fact varying degrees — making it softer (and more difficult) as it gets nearer the surface. These therefore, are the areas which require caution when drilling and also the ones that could contain a larger amount of water.

This weathering can make the rock as soft and loose as flour and it is here that some drillers have come to grief when thinking that all igneous rock areas should always be drilled with a hammer. They shouldn't. Try rotary first, until you have penetrated the softer sections. Sometimes you even have to revert to rotary after a hammer if a soft layer is encountered — you could cavitate.

Drilling tools? You guessed it: a drag bit, usually with foam/polymer flushing (stabilizing the soft rocks) not only for the soft layers but for the occasional bands of quartz pebbles that a drag bit can handle fairly well. A hammer is essential for the hard areas and, just as important, a grinder for your bits — there can be a high degree of abrasion.

Are igneous rocks easy drilling? Not a bit of it. Just the mention of the word basalt can strike terror into the hearts of drillers.

Volcanics is a better name for these basaltic rocks. Basalt is the result of an ancient lava flow that has cooled rapidly in the atmosphere, forming a cap, or layer, or several caps or layers over the years. These layers can vary in thickness from stringers to hundreds of feet. They can be massive (dense) or porous (vesicular) and can sandwich a hotch-potch of anything like soft volcanic ash, pumice or voids so you don't know where you are at until you get to bedrock, which can be just about anything. This is best illustrated with a story.

There were these oil rigs in the Far East, drilling geothermal wells in volcanics. They had to set 20″ surface casing in a 26″ hole to 300 feet. Their problem — massive mud losses in the layers of vesicular basalt and total loss in the soft layers in between where their huge mud pumps were washing out the soft materials — the term is "washout".

They had to cement off the wash-outs, which occurred up to 90 — yes 90 — times, in the 300 feet; it was costing them an average of one and a half million dollars for this section because of the considerable amount of time it took to complete this task as well as materials — on average 45 days.

We flew two men, (twenty-four hours' working), a foam pump and a bunch of foam and polymers to the site and rented a battered old compressor off the side of the road; when they arrived and set up next to this enormous rig, the crew just laughed and laughed. 48 hours and ten minutes later they smiled — with pleasure — because we had got them out of an expensive mess.

The job should have been completed in a few hours, but they didn't have a hammer available for the hard basalt.

So — there you are — volcanics. Groan no more if you go about it the right way. A hammer for the rock, foam to lift cuttings, polymers to stabilise the soft bits in between and just enough (hammer and rotary!) air to power them round without disturbing the formations.

To summarise igneous rocks. With the exception of volcanics, they are nothing that your basic rotary tools and hammer can't handle (with foam/polymer of course). The volcanics requirement is the same in tooling, but the cap can overlay soft materials, including water bearing running sands. So you will need a mud pump as well if only as a stand-by (if you ain't got it — you will need it — remember?).

But — don't forget your bit grinder.

Metamorphic rocks

Metamorphosis is when a rock is changed by heat and pressure, and is not to be confused with changes brought about by compaction (in the latter clay becomes shale, for instance).

For example:-

Igneous rocks become gneiss or schist
Sandstones become quartzites
Shales become slates
Limestone Chalk becomes marble

These rocks can usually be recognised by the fact that they will split at right angles to the direction of the pressure so, if the pressure was downwards, then rocks will have splitting planes in a horizontal direction, with all the appearances of a stack of sheets of paper.

Metamorphosed igneous rocks (gneiss and schist) have another trade-mark in that they can contain a considerable amount of mica, and will tend to shine. Gneiss will be found to be coarse grained, while the schist will be fine grained.

The bad news is that these formations are unlikely to yield much water. In fact they are probably the least fruitful of all three categories.

Drilling-wise, they can be relatively uneventful if you have your basic systems (rotary/hammer/foam/polymer) on site; you should cope well, although beware of water losses.

Sedimentary formations

In these water can be abundant, as can the drilling problems for the unwary driller.

There are four main types and we will list them along with their sub-headings.

1) *Arenaceous*

GRIT. Quartz grains, very coarse, large and may have small pebbles present. Generally a massive strata coarsely bedded. Nasty but nice for water. Rotary mud drilling is recommended or, if the hole is of a small diameter, try simultaneous drilling and casing. If in solid form can be very hard and extremely abrasive — therefore a hammer.

SANDSTONE. Finer grained than grit, with the grains cemented together. Very nice, and can be a prolific producer of water. We have had wonderful results with rotary drilling but there have been times when it got a bit hard and we went over to a hammer.

FLAGSTONE. A bedded sandstone which will split in layers. Good drilling and some water. Mostly hammer.

2) *Argillaceous*

SHALE. Thin-layered compacted clay which will split into thin sheets. Little water but what there is can swell (slough) the formations into the hole. A nasty condition, overcome by the correct polymer. Shale will be interspersed with something else, but as its parentage is clay you should be thinking of rotary drilling with the essential polymers.

MUDSTONE. Like shale but no splitting. Good drilling, especially with drag bit. Watch discharge velocities at the bit. Erosion is possible.

FIRECLAY. A mixture of clay and silica with fossils. Can get very sticky and swell if you do not use correct mud/foam additives and can be very soft. Will not actually contain much water. It is clay, so rotary drilling please.

3) *Calcareous*

LIMESTONE. There all sorts of limestones. Just to mention a few: oolitic; carbonaceous; dolomitic; chalk. These can be all different colours from white, to grey, to red, to all sorts.

They can be soft through to extremely hard, and they will be cracked or even cavernous. They can be abrasive or non-abrasive, but they will almost certainly produce reasonable amounts of water.

You will need rotary/hammer tooling and your system will be subject to lost circulation and water, so foam/polymer is essential. Do you remember our earlier story about massive lost circulation in Libya? Well that was in dolomitic limestone.

4) *Organic*

This is really the remains of plant life that has undergone chemical and pressure changes resulting in peat, lignite and coal. You could find a lot of water hanging about, although not necessarily in it — good rotary work with mud or foam/polymer.

Watch the sandstones for abrasion and, at all times, take your bit grinder with you.

Let us summarise with a sort of check list and at the same time — in more understandable geological headings — give a slightly different approach to the above.

Formation	Problems	Rotary/ Foam	Rotary Mud	Hammer
Loose sand/ gravel	Running sand	no	yes	no
Boulders	Collapse	no	no	yes
Clay/silt	Erosion/ Swelling	yes	yes	no
Shale	Sloughing	yes	yes	no
Sandstone	Collapse	yes	yes	yes
Chalk	Flints	yes	yes	no
Limestone	Lost circulation	yes	no	yes
Basalts	Lost circulation	yes	no	yes
Metamorphics	Poor water	no	no	yes
Igneous rocks	Hardness	no	no	yes
Overburden*	Collapse/ Erasion	yes	yes	no

* Re overburden. Please note that this almost always overlies rocky formations and must be considered as a separate entity and treated with great respect. You cannot expect to use the same tools in a hole from top to bottom.

The above list of suggested guides to drilling applications is based on the understanding that drilling systems for the specific conditions written about in this series of books are used and used diligently.

9 Handy Formulae

Foreword to Book Nine

The formulae which follow are as important as anything else because they allow the uninitiated to construct a "picture" of what happens in the field without deep field knowledge. If the financial personnel in the company can see their way to making a few extra "cents" by becoming efficient, that is a great help to people in the field — let them work it out and the field people fill in the "nuts and bolts" of the job.

Handy Formulae

These formulae are guides towards efficiency but, as nature is unpredictable, should be treated with caution.

The rig

Wouldn't it be wonderful if we could take a rig which has a hoisting capacity of ten tons and happily drill away until we have ten tons of tools on the hook, with the sure knowledge that you could pull it out at the end of the day. You can't can you, because "mother nature doesn't give up her riches lightly" and she is going to set all sorts of traps for you.

Imagine the said ten tons (in fact 9,999 kilos) is in the hole and "herself" decides to drop a little stone on top of your bit and it lodges there. You only have one kilo of "overpull" left, which is no good whatsoever, so you've lost the tools, more than likely. No, you drill safely and say that you are going to leave a margin of error which is a percentage of the total capacity as an "overpull" — we say you need to leave twenty-five percent. So:-

Take the capacity of the rig which, here, for argument's sake, is ten tons. Therefore if you reduce this to allow for the safety factor of twenty-five percent, you are left with seven point five tons as the "usable" capacity.

Then you have, say, two tons of drill collars which leaves five point five tons (5,500 kilos). Your drill-pipe weighs 20 kilos per metre, therefore your rig capacity is:-

$$\frac{5500}{20} = 275 \text{ metres } \textit{plus the length of the drill collars.}$$

Rotary torque

A similar thing applies here. You cannot rate your rotary head (table) at maximum torque because the offending stone can play havoc at your bit, denying you the one hope you had of getting out — the ability to turn the bit.

Let us stop here and introduce another one of our sayings:-

"If you can't go up go down".

What does that mean? It means, if your bit is stuck and won't come up try going down again — and keep trying — up and down — it works wonders.

Back to torque. You have to have a margin of safety and we use the same figure — twenty-five percent.

Torque — rotary drilling (figure 9-154A)

We know that it takes one hundred and fifty pounds/feet of torque to effectively turn one inch of diameter (loaded as per later). So, if we have a ten inch bit, that is one thousand five hundred pounds/feet of torque plus one third safety factor.

But, you say, one third is 33.333% when we only need 25%. That is correct because, you must add 33.333% on to get 25% off. Example — 100 plus 33.333 is 133.333, less 25% is 99.99999 whereas 100 plus 25(%) is 125 less 25% = 93.75 — work it out for yourself.

Anyway, a 10″ bit needs $10 \times 150 \times 1.33 = 1995$ pounds/feet of torque including safety.

Torque — hammers

Rotary torque for a hammer is similar because, whilst the bit is rigid (therefore should require more torque?), it is relatively lightly loaded.

Torque — augers

The torque calculation for auger drilling is similar to rotary drilling at 150 lbs/ft per inch of bit diameter, but you must add to this the extra torque needed to overcome the friction of the flights against the sides of the hole and the weight of the carried cuttings, so we add two pounds/feet per inch of diameter per foot run. So:-

A ten inch continuous flight auger to 100 feet would require 10 (inches) × 150 (lbs/ft) × 1.33 (safety factor) + 10 (inches) × 2 (lbs/ft) × 100 (feet depth) x 1.33 (safety factor) which equals 4655 lbs/ft.

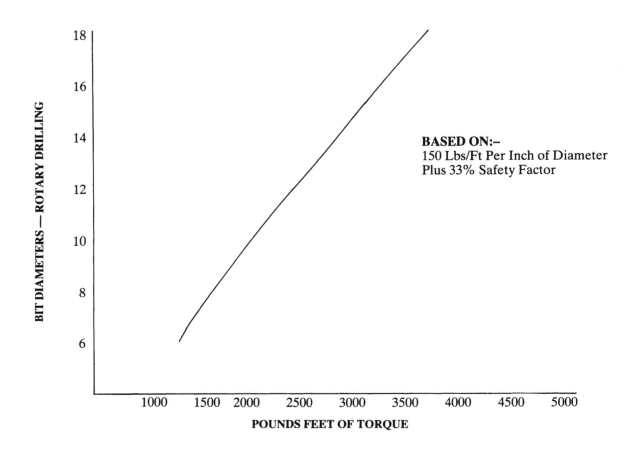

BASED ON:–
150 Lbs/Ft Per Inch of Diameter
Plus 33% Safety Factor

(Y-axis: BIT DIAMETERS — ROTARY DRILLING)
(X-axis: POUNDS FEET OF TORQUE)

Weight on bit (collar) for rotary drilling

We like to see 1000 pounds per inch of diameter — minimum — wherever possible for a rockbit and half of that (500 pounds per inch) for a drag bit in the harder formations, reducing for the softer to maybe half that. But never use pulldown, only drill collars (the exception being at the top of the hole where prudent pulldown is permissible).

If you are rotary drilling hard formations, then you either have to suffer the consequences of poor performance (if you cannot apply the above minimum), or get yourself a hammer or a bigger rig — QED.

Weight on bit — down-the-hole hammer

Really quite critical. If there is not enough weight on, it will tend to bounce up and down around the bit shank and break it (or something else). Too much weight and you will scrub out tungsten carbide inserts as well as restricting the overall efficiency of the tool.

All that is required is to overcome the air pressure in the hammer and to maintain that weight throughout the section to be drilled, even when adding (or subtracting) drill-pipe. Never use a drill collar, you will break something. How do we work it out?

You *need to know* the surface area of the

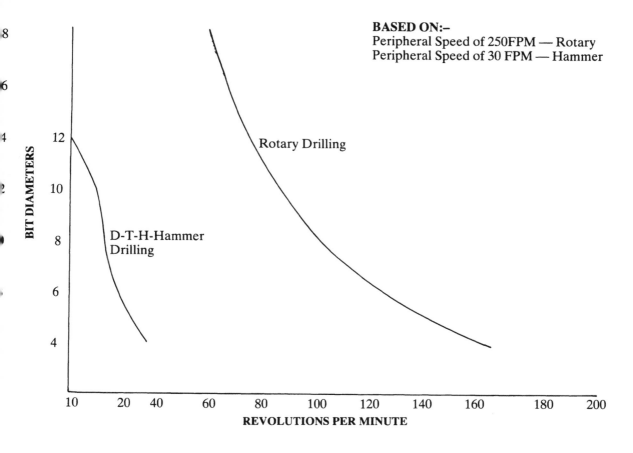

tary   Hammer

BASED ON:–
Peripheral Speed of 250FPM — Rotary
Peripheral Speed of 30 FPM — Hammer

Rotary Drilling

D-T-H-Hammer
Drilling

BIT DIAMETERS

REVOLUTIONS PER MINUTE

hammer piston here, for instance, ten square inches. You then multiply that by the pressure of your compressor, say 200 psi so, you multiply ten square inches by 200 psi and you get 2000 pounds weight on bit, then add a bit to overbalance to, say, 2200 pounds total.

Make some allowance for pressure losses through the drill-pipe as depths increase, and also allow for flotation of the string in water. Here, listen to the hammer (drill with your ears), and when you think it correctly loaded (and performing well), compare the sound as you go down.

As you don't use drill collars, a prudent amount of pulldown is permissible at the top of the hole if the ground is hard, but counterbalance that extra weight of tools as you go down.

Rotary speeds (figure 9-155A)

We have said a great deal about these elsewhere. Suffice to say here that they be looked at in their context e.g. rotary or hammer.

In both cases speed is looked at as peripheral; that is the time it takes for a point on the circumference of the bit to travel to the same point again — in feet per minute.

Speed — rotary drilling

As a general rule the peripheral speed for rotary drilling is 250 feet per minute so, to find the rpm for a 10″ bit (for example), the formula is:-

$$\frac{250}{Pi \times d} = \text{revolutions per minute (rpm) where:-}$$

250 = 250 feet per minute peripheral speed
Pi = 3.14
d = diameter of bit in inches.

Therefore for a 10″ bit:-

$$\frac{250}{(3.14 \times 10 \div 12 \text{ (inches)})} = \frac{250}{2.61} = 96 \text{ rpm}$$

Rotational speed — down-the-hole hammer

The peripheral speed we work to for the hammers is 30 feet per minute, but remember, once you have got to the guide speed, alter the rotation slightly up and down and watch your cuttings — the point where the cuttings are largest is the best speed. This also applies to rotary drilling.

So, for a ten inch hammer bit:-

$$\frac{30}{(3.14 \times 10 \div 12)} = \frac{30}{2.61} = 12 \text{ rpm.}$$

Calculation of annular velocities (figure 9-157A)

Here we will deal with pure air, water and mud circulation. We will not deal with foam/polymer because it can move at almost any speed and still be efficient. Remember, you must still have enough air to operate the hammer, albeit with a blank choke.

We will look at calculation for work with existing pumps, as well as looking at pumps for future work.

Annular velocity — water

The minimum annular velocity for water is 120 feet per minute. Why? Because, stones dropped in water will settle at an average rate of 60 fpm and we double this. The maximum we like to see is 300 fpm and even then you are in danger of eroding the looser formations. Be very careful here because you could change the flow from the desired laminar flow to the dangerous turbulent flow. Check it out.

As almost all mud pumps are rated in United States gallons per minute (USGPM) we will use these figures. However, for imperial gallons per minute the factor of 24.74 (USGPM) should be substituted with 29.4 (IGPM).

To find the annular velocity of water from a given mud pump, given hole size and given drill-pipe size, not forgetting to use the biggest annulus in the hole which, for instance, could be the diameter of the kelly inside, say, surface casing, the formula is:-

$$\frac{\text{Factor} \times \text{capacity}}{(D \text{ squared} - d \text{ squared})} = \text{velocity.}$$

Where:-

Factor = 24.75 (USGPM)
Capacity = mud pump capacity in USGPM
D = diameter of hole or surface casing
d = diameter of drill-pipe/kelly.

If your mud pump is rated at 250 USGPM, your surface casing is 10″ inside diameter and you are using 4″ drill-pipe:-

$$\frac{24.75 \times 250}{100 - 16} = \frac{6187.50}{84} = 73.66 \text{ feet per minute.}$$

This is far too slow and you should not start the hole — remember — "if you can't clean the hole don't start".

Let us now find out how big our pump should be. The formula is:-

$$\frac{(D \text{ squared} - d \text{ squared}) \times \text{velocity}}{\text{factor}} = \text{capacity USGPM}$$

156

Fig 9-157A

A GUIDE TO ANNULAR VELOCITIES

	Imperial Gallons Per Minute	United States Gallons Per Minute	Cubic Feet Per Minute	IGPM	US GPM	CFM	IGPM	US GPM	CFM	IGPM	US GPM	CFM
5"										113	134	772
4½"							52	62	360	124	147	850
3½"	19	23	131	30	36	208	71	85	490	143	170	981
2⅞"	29	34	196	40	47	274	81	96	556	152	181	1046
2⅝"	32	38	219	43	51	296	84	100	578	156	185	1069
2⅜"	35	41	240	46	55	317	87	103	599			
	4½"			5"			6½"			8½"		

DRILL PIPE DIAMETER (rows) — HOLE DIAMETER (columns)

A GUIDE TO ANNULAR VELOCITIES

	Imperial Gallons Per Minute	United States Gallons Per Minute	Cubic Feet Per Minute	IGPM	US GPM	CFM	IGPM	US GPM	CFM	IGPM	US GPM	CFM
5"	173	205	1185	298	354	2044	458	545	3148	670	795	4598
4½"	184	219	1263	309	367	2121	470	558	3226	681	809	4676
3½"	203	241	1394	328	390	2252						
2⅞"	213	252	1459									
2⅝"												
2⅜"												
	9⅞"			12¼"			14¾"			17½"		

DRILL PIPE DIAMETER (rows) — HOLE DIAMETER (columns)

LEGEND:
1) Mud volumes calculated on a minimum uphole velocity of 70 feet per min.
2) Air volumes calculated on a minimum uphole velocity of 3000 feet per min.
3) No allowance is made for, say, surface casing that has a greater diameter.
4) Pumps should have an excess of air/mud available over and above that shown.
5) These volumes are a guide and should be treated as such.

Therefore for the same set of circumstances:-

$$\frac{84 \times 120 \text{ (fpm min.)}}{24.75} = 407 \text{ USGPM required.}$$

Annular velocity — mud

Now, if you are using mud, the required annular velocity is lower than water as the cuttings are suspended better in the mud. An absolute minimum annular velocity with, say, a mud viscosity of 40 secs., would be 70 feet per minute.

Let us return to the same set of circumstances as with water and you will see that, at 70 fpm, our pump is big enough (73.66 fpm calculated). All well and good but that is near minimum — OK?

Annular velocity — pure air

With pure air the minimum annular velocity is 3000 fpm and the maximum 5000 fpm. The factor is 183.5, so let's have another go at the above, assuming that our 10″ hammer takes 450 cfm (cubic feet of minute) when drilling. Do not use the "blowing" rate as that is intermittent: we also assume that this is all the air you have for rotary drilling:-

$$\frac{183.5 \times 450}{84} = \frac{82575}{84} = 983 \text{ fpm velocity.}$$

This is hopelessly slow, so, use foam/polymer or, if not available, bigger drill-pipe, which can easily be calculated from the above formula. Now, assuming we are rotary drilling, we could get away with a bigger compressor, which we couldn't with a hammer as it will not absorb any more air. Our compressor would have to be:-

$$\frac{84 \times 3000}{183.5} = 1374 \text{ cfm to achieve min. 3000 fpm velocity.}$$

* * *

The Bottom Line

We will now look at how much it costs to run your rig, because you will then see how expensive it is for the rig to stop working for even one hour. Remember — "when it's turning it's earning". There are times, of course, when that saying doesn't apply: When, for instance, you are pump testing, you are not "turning" but "earning" by the hour.

Costings are divided into two clear parts: fixed costs and running costs. The first are ongoing, when nothing is happening. For instance, when you are fishing, which does not earn a cent, but increases those fixed costs.

Running costs — if you look after your tools they can be minimal. But, start scrubbing out bits and these costs will shoot up.

Now for the bases (formulae) we work to:-

Costing formula (estimate)

FIXED COSTS

A   Depreciation. Total cost of capital equipment divided by 10,000 hours gives depreciation per hour.
B   Maintenance. Take 5% of depreciation hourly cost. This is maintenance cost per hour.
C   Labour. Take actual cost per hour.
D   Overheads. Take same figure as labour per hour, which gives overheads per hour (a rule of thumb).
E   Fuel cost per hour.
F   Mud/foam cost per hour — say US$10.00.

Running costs per metre

1   Take drill string cost and divide by 20,000 metres.
2   Rock bit cost divided by 300 metres.
3   Down-the-hole hammer cost divided by 3,000 metres.
4   Down-the-hole hammer bit cost divided by 300 metres.

To estimate cost per metre for drilling, take total cost per hour (fixed cost) and divide by estimated drilling rate then add in running cost per metre, selecting type of drilling (hammer or rotary) this gives your actual cost per metre.

For standing time take total fixed cost per hour.

For casing, take landed cost of casing plus handling charge then add standing time rate.

Remember — when the rig is stopped, fixed costs go on and the slower the drilling rate the greater the cost.

Let us have a go at working out an example, first making some assumptions which, here, are based on a random set of figures for which you should substitute your own accurate figures. The result here will be a cost figure therefore will not include any profits.

Assumptions

A Value of all capital equipment (incl trucks etc.) US$500,000.00
B Maintenance at 5% of capital cost.
C Overall labour cost is $25.00 per hour.
D Overheads as per labour cost at $25.00 per hour.
E Fuel cost at $30.00 per hour.
F Mud/foam $10.00 per hour.
1 Drill string valued at $100,000.00.
2 Value of rockbit $2,000.00.
3 Value of hammer $15,000.00.
4 Value of hammer bit $2,000.00.

Fuel and mud are included in fixed costs because they have to be purchased and, at times, are used when not drilling e.g. when casing or cleaning the hole etc.

Fixed cost per hour

Depreciation $\dfrac{\$500,000.00}{10,000 \text{ hrs.}}$ = $50.00 per hour.

Maintenance 5% of $50.00 $2.50 per hour.

Labour per hour $25.00 per hour.

Overheads per hour $25.00 per hour.

Fuel cost per hour $30.00 per hour.

Mud/foam cost per hour $10.00 per hour.

Total fixed cost per hour. **$142.50 per hour.**

Running cost per metre

Drill string	$\dfrac{\$100,000.00}{20,000 \text{ mtrs}}$	$5.00 per metre
Rockbit	$\dfrac{\$2,000.00}{300 \text{ mtrs.}}$	$6.66 per metre
Hammer	$\dfrac{\$15,000.00}{3,000 \text{ mtrs.}}$	$5.00 per metre.
Hammer bit	$\dfrac{\$2,000.00}{300 \text{ mtrs.}}$	$6.66 per metre.

You will note that it costs, repeat, *costs* $142.50 per hour to do nothing and this figure, multiplied by the number of hours lost, will come off your profits for the job. Remember that.

To arrive at an overall cost per foot (without profits) we have to presume a drilling rate which we will guess at an overall 3 metres per hour for 10 metres of overburden and 110 metres of rock. Your casing cost and development etc., must be added, but as we don't have any figures, use the formula above and do it yourself.

Fixed cost per metre is $142.50 ÷ 3 metres per hour which equals $47.50 per metre — OK?

For the overburden, we used a rockbit for 10 metres multiplied by $6.66 per metre. Add in drill string cost for 120 metres × $5.00.

We drilled 110 metres of rock with the hammer × $6.66 for the bit and $5.00 for the hammer. Now, what have we got for a cost for drilling the hole — only drilling and *at cost*?

Fixed costs per metre.
$47.50 × 120 metres $5,700.00

Drill string.
$5.00 × 120 metres $600.00

Rockbit.
$6.66 × 10 metres $66.60

Hammer.
$5.00 × 110 metres $550.00

Hammer bit.
$6.66 × 110 metres $732.60

Drilling cost for hole $7,649.20

An expensive business isn't it? This is a simple
formula and can withstand a lot of pressure if you
remember you must add all other functions,
based on the formula, and your profits.

10 Incidentals

Book Ten — Incidentals

A SUMMARY OF SAYINGS

A GLOSSARY OF TERMS

A Summary of Sayings

Throughout the foregoing books a number of sayings which apply to drilling have arisen which might need a little expansion and, you never know, could well apply to other things.

We will talk mainly about what happens at the well-site, because we automatically assume that your administration will do everything possible to make your job smooth and that between you, you applied all the formulae contained in Book Nine plus all the other information to arrive at the well-site in good fettle.

"There are no short cuts in drilling"

Let us discuss a few reasons why we made up this saying:-

1) Buying second quality bits is a false short cut because, if you look at the formula concerned with costings in Book Nine, you will see how significant the cost per metre for a bit is. A rockbit that costs $2,000.00 and gives you 300 metres costs $6.66 per metre but, let us say you got a second quality bit for $1,500.00 and it only gave you 150 metres, your cost per foot is $10.00 per metre, and we are being kind to the second quality bit by suggesting it will achieve half the performance of a good bit.
2) You have your nice rig but no drill collars. You push the hole down with the rig and then can't get your casing in or, as happened here just recently, a hole was five feet out in 300 feet of drilling, whereas the limit was one foot in three hundred. The contractor had to redrill.
3) A contractor we know needed some empty drums to get his diesel fuel in. He got them OK and was too lazy to clean them out, with the result that he "gummed-up" his fuel lines, losing him days of drilling time. Just look at cost per hour in Book Nine.

"When it's turning it's earning"

There is an American version of this — "cram that iron in the hole and turn it to the right" — but we don't like that one so much.

All this means is that all the while the bit is turning in the hole (drilling of course) it is earning money (or making water for charities). When it stops for anything other than paying work, the fixed costs go on being used up. These are *deductible from profits*.

Let us use the cheap bit case again. OK, the bit was cheap and there isn't much apparent difference between $6.66 and $10.00 per metre (it is quite significant really — work it out for yourself). But you have to make one more round trip and that takes you three hours, including getting the bit off the string. That is $3 \times \$142.50 = \$472.50$, plus the difference in the cost per metre for the two bits, another $500.00 or so. That means it has cost you nearly five hundred dollars more than the price of a first quality bit to buy a cheap bit.

"The simpler the better"

This has to be looked at carefully. Remember how we went into, for instance, the difference between a simple gear type hydraulic pump and an expensive variable flow unit? In terms of maintenance at least, the simple type will save you time, therefore money, and an enormous amount of frustration if you are remote from good servicing facilities.

SIMPLE DOES NOT MEAN CHEAP!

"A happy crew will make money for you"

And that is a fact. A crew that is working in "the bush" without adequate food, accommodation and equipment cannot possibly give you their undivided attention.

"If it's strong enough it's good enough"

In the context in which it was used this is very true. Take for instance, pipe wrenches. You can buy excellent ones from Europe and the United States of America but there are copies on the market from Eastern Countries which should not be used under any circumstances for breaking out drill-pipe — they are not strong enough and injury to personnel when they break, which they do very easily, is not uncommon.

"A bent hole is a spent hole"

We have already touched on this above but there is no harm in repeating it — use drill collars for rotary drilling and watch your weight on bit. On the latter point, this is very important when it comes to down-the-hole hammers. Just have another look at bit and hammer costs in Book Nine.

Not only can't you get your casing in without difficulty (sometimes not at all) but your tools scrape the side of the hole, causing unnecessary wear to them and untold damage to your hole. The problems are endless — don't forget what we said above: the client can decide not to accept a bent hole, and where are you then?

"The thinking starts here"

Do you remember what Mr. Harry Truman, an ex-president of the United States said — "The buck stops here"? Well, as soon as your bit disappears below surface you've got to think about that hole all the time, because any wrong decisions made by you can cost a lot of money and you can cause yourself lots of trouble with "the boss". Why?, because "the buck stops with him".

Listen, look and feel your way to the end of the project. Everything is on your side if you obey the rules — and you have a lot of allies.

"When the engine blows coal there's trouble in the hole"

With regard to what have we just said about "allies", this is a typical example. Your engines (rig and pump) will take the strain of undue stress being put on them by adverse hole conditions; the exhaust gases will turn "as black as coal".

Just supposing you are having a cup of well earned tea and your assistant is at the controls and is not watching the torque gauge (you are within ear/eye shot aren't you driller? You must always be). You will be able to see (black exhaust) and hear (the dying revs.) that there is "trouble in the hole".

"Never drill faster than you can clean the hole"

This is so fundamental that it hardly needs explanation — but we will.

If you drill faster than your air/mud/foam can clean the cuttings from the hole you will bury the bit and more than likely get stuck and even worse, lose your tools and the hole. You can also make a mud ring, which will result in the same problems.

"An expert fisherman is not necessarily a good driller"

If the drillers spend too much time fishing that means they either have a poor set of tools (even that is their own fault) or they cannot work within the capacity of the equipment — all of it.

There was an old man who spent his life fishing. I don't think he was ever a driller, only an assistant who had a knack of visualising incidents in a hole.

"Good mud control makes lots of hole"

Keep your hole clean and your mud free of solids, then your pumps, mud lines etc. will only suffer normal wear and tear, giving you that much more time for drilling (none spent on renewing rapidly worn parts) — this is done by good mud control.

"If you've got 'em you won't need 'em. If you haven't you will"

This is superstition more than anything else but beware of it.

Our point here was aimed at fishing tools but it applies to other things as well — don't tempt providence.

A Glossary of Terms

A.B.S. A plastic (Acrylinitrite Butediene Styrene) used in some casing manufacture.

ABRASION. A term describing wear caused by frictional contact.

ADAPTORS. See Sub.

AGITATOR, MUD. See Mud Agitator.

AIR DRILLING. The use of compressed air for cleaning the hole and cooling the bit.

AIR LIFT DRILLING. A reverse circulation drilling system using compressed air to lighten the column of water inside the drill-pipe.

AIR LIFT. A method of pumping water by passing air down one of two tubes, the water then passing up the hole inside the outer tube.

AIR-LINE OILER. Fitted into the air-line, allowing a controlled amount of lubricating oil to enter the airstream for, say, hammer lubrication.

API. Abbreviation for American Petroleum Institute.

ASSISTANT DRILLER. Will relieve the driller and perform duties around the rig such as mud supervision etc.

ATTAPULGITE. A clay used in mud drilling, which is sympathetic to saline water.

AUGERS. A self-cleaning drilling tool based on the Archimedean Screw.

BACK OFF. To unscrew a thread.

BACK PRESSURE. A resistance to a pressure, or that pressure caused by frictional losses.

BACK-UP (ANCHOR). Resistance to an applied force.

BACKFILL. Filling a hole or annulus with solid matter.

BAIL. An inverted "U" shaped lifting device fitted to, say, a lifting plug. Also see Bailer.

BAIL. In cable tool drilling, the process of cleaning the hole with a bailer.

BAILER. A tube with a check valve (dart valve) at the bottom and a bail at the top which is lowered into a well for cleaning and/or development.

BAKER VALVE. Proprietary check (or float) valve.

BALL VALVE. A non-return (check) valve with a ball as the valve.

BARAFOS. A proprietary name for a polyphosphate used for cleaning out mud residue from a well during development.

BARITES. A heavy mud used for the control of formation pressures. Has a specific gravity in excess of 4.0.

BARREL. A container with a usable volume of 42 U.S. Gallons.

BELL BOX. A tubular fishing tool employing a latch or latches.

BELL-NIPPLE. A short pipe positioned on top of the hole with an outlet angled downwards through which mud from the well is discharged onto the, say, shale shaker.

BELL TAP. A female fishing tool with hardened cutting threads on the inside.

BENTONITE. A clay based drilling mud.

BIT BREAKER. A device shaped to fit around a bit to help in backing it off.

BIT CHECK-VALVE. A non-return valve directly behind the bit to prevent washback. An essential part of the drill string.

BIT GAUGE. Used to measure bit wear and, often, bit clearance angle.

BIT GRINDER. A tool for "dressing" (grinding) bits. Can be a simple grinding wheel or, with down-the-hole hammer bits, specially devised cups in the profile of the tungsten-carbide buttons.

BIT NOZZLE. Nozzles fitted to jet type rockbits.

BIT SHANK. That part of a bit which connects the cutting section to the tool above above it — say, a sub. or hammer.

BIT THIRDS. The three sections of a rockbit before welding. Often used in the manufacture of hole-openers.

BIT. The tool that actually cuts the hole (or crushes — etc. see text).

BITCH. A plate-like device used to hold augers and upset pipe in the table.

BLOOEY LINE. A diverter pipe taking cuttings or pressure away from the well.

BLOW-OUT PREVENTER. A device used to control formation pressures preventing excessive and potentially dangerous discharge from the well.

BLOW-OUT. A sudden escape from the well caused by excessive formation pressures.

BORE. The diameter of a piston in a pump/engine.

BOREHOLE. See Water Well.

BOREWELL. See Water Well.

BOX. Determines a female thread — thus box thread.

BREAK-OUT TONGS. Wrap-around wrenches for unscrewing drill-pipe, collars etc.

BREAKING OUT. Unscrewing, say, drill-pipe.

BRIDGE PLUG. A plug set at a given point in a well upon which cementing is to be done.

BRIDGE. An obstruction in a hole.

BRIDGE. The act of locking together.

BSI. The British Standards Institute.

BULL REEL. The largest winch on a cable tool rig off which drilling is done.

BULLDOG SPEAR. Non-releasing recovery tool for use with casing and the like.

BUMPER JAR. Used to release tight tools in cable tool drilling.

BUOYANCY. A drill string floats somewhat in mud, therefore, the weight on bit is reduced. This is known as the buoyancy factor.

BUTTON BITS. A percussive bit with cone or chisel-shaped slugs of tungsten carbide.

BY-PASS. A system of valves on a mud-pump discharge allowing mud direction to be changed, e.g. from the standpipe to the mud mixer.

CABLE CLAMP. A threaded "U" bolt with cross member to clamp two wire ropes. The cross member must always be placed over the longer of the two wires to be clamped.

CABLE PERCUSSION RIG. Drills by dropping a weight (bit) suspended from a wire rope.

CABLE TOOL. See Cable Percussion Rig.

CABLE TOOLS. Drilling tools used with a cable percussion rig.

CAKE. Deposited on the wall of the hole by clay based drilling fluids.

CALCIUM CHLORIDE. A cement accelerator.

CALF REEL. The smaller of the two main winches on a cable tool rig.

CALIFORNIAN CHISEL. Cable tool bit with sloping shoulders.

CALIPER. Sprung device used for measuring diameters.

CAP. (To Cap) To seal or close off a well.

CAPACITY. The power of holding — rig capacity is its rated lift.

CARBIDE INSERT. Pieces of tungsten carbide fixed to a body to form a cutting tool.

CAROUSEL LOADER. A method of stacking drill pipe in a revolving device in the mast of a top drive rig, thus achieving semi-automatic make-up.

CASING. A steel or plastic tube put into a hole for support, abstraction, control of mud losses etc.

CASING CLAMP. A bolt-up pair of semi-circular shaped supports, with arms, to hold casing.

CASING COUPLING. A short piece of tube box/box or pin/pin-threaded to connect two pieces of casing.

CASING CUTTER. A tool used for cutting casing in situ.

CASING DRIVE HEAD. Fitted to top of casing when driving by percussive means.

CASING HANGERS. A device by which casing can be set inhole on drill-pipe and, with the hanger actuated, allows the drill-pipe to be released.

CASING JACKS. Used to pull stuck casing.

CASING PERFORATOR. Used to perforate casing in situ.

CASING SHOE. Fitted to the bottom of casing for "drilling in" in one way or another. Used when driving (drive shoe-bevelled) or rotating (serrated, hard faced, diamond).

CATHEAD. An auxiliary winch used with a slip rope.

CAVING(S). The act of collapse in a hole or the particles therein.

CELLAR. An excavation under a mobile rig to accept, say, a blow-out-preventer when there is insufficient room normally.

CEMENTING. The act of introducing cement into an annulus or hole.

CEMENTING CAP (HEAD). A device fitted to the top of casing to be cemented in. It will have a connection to which to attach the cementing lines and a pressure gauge.

CEMENTING LINE. Connects the cementing pump to the casing cap.

CEMENTING SHOE. A casing shoe especially designed for cementing, it can contain a drillable non-return valve in a drillable matrix.

CENTRALISER. A device attached to the outside of casing, thus keeping it central inhole. Essential when cementing or gravel packing.

CENTRE ROPE SPEAR. A barbed fishing tool for retrieving wire-line from inhole.

CENTRIFUGAL PUMP. A pump with an impeller rather than pistons.

CFM. Cubic feet per minute of compressed air.

CHAIN FEED. Where the up and down motion of the tools is controlled by chains.

CHAIN TONGS. A wrench actuated by a jaw and chain. Used mainly for casing.

CHECK VALVE. A non-return valve. Can be either ball or dart.

CHISEL BIT. A percussive bit with a single cutting edge across the full diameter.

CHOKE. Used to control the amount of free air passing through a down-the-hole hammer.

CHOKE. Usually in sets, attached to the blow-out preventer stack and used to over-balance the well in a pressure situation.

CHURN DRILL. See Cable Percussion Rig.

CIRCULATE. A flushing system where fluid passes to surface, is cleaned, then pumped back into the hole — a "circular" motion.

CLACK VALVE. A circular valve hinged at one side, allowing the ingress only of fluids.

CLEARANCE ANGLE. The angle from the outward-most point of the bit from vertical. Usually around four degrees.

CMC. Carboxy Methyl Cellulose.

CONDUCTOR PIPE. A short length of casing used in the top of a well.

CONNECTION. Making up a new joint of drill-pipe/rod/collar.

CORE. A cylindrical sample of formations drilled by a core barrel/bit.

CORE BARREL HEAD. The top of a core barrel which is attached to the drill rod/pipe.

CORE BARREL. A tube or tubes which, with a core bit, produces a core sample.

CORE BIT. An annular bit for cutting cores. Can be diamond or tungsten carbide inset.

CORE BOX. A box sectioned to accommodate core samples.

CORE CATCHER AND SPRING. A toothed spring inside a tapered cylinder (core catcher) fitted behind the bit. The taper closes the spring around the core on withdrawal.

CORE DRILLING. The process of obtaining core samples.

CORE EXTRACTOR. A device for removing core samples from the core barrel.

CORE RECOVERY. The amount of core recovered relative to the amount drilled (%).

COUPLING. A double threaded device to connect two threaded tubulars. Mainly used to describe a "casing coupling".

CROSS-OVER SUB. See Substitute.

CROWD. See pulldown.

CROWN BLOCK. The sheaves and sheave mounting at the top of a derrick.

CROWN. See core bit.

CRUCIFORM BIT. A percussive bit with four cutting edges at 90 degrees to each other.

CUTTINGS. The particles of the formations cut by the bit.

D SHACKLE. A "D" shaped device where the straight line of the "D" is a screw-in pin. Used for the connection of drilling tools etc. to a wire line.

DART VALVE. A valve with machined faces which has a protruding striker plate to open the valve, thus allowing fluid to pass in one direction only — see bailer.

DEADLINE. The end of a wire-line fixed to the derrick which is motionless when the travelling block rises and falls.

DEADMAN. A fixed object to which a wire-line is attached, acting as an anchor.

DEBRIS. Alien matter in-hole.

DEFLECTION. An intentional change of direction in the drilling of a well.

DEFLECTION BIT. A special bit for use in deflection drilling.

DERRICK. The tower, or mast, of a drilling rig.

DERRICK MAN. Will work the derrick, placing the upper end of the drill-pipe/drill collar stands into the finger board.

DE-SANDER. A device comprising a pump and one or more cones for removing sand from drilling mud.

DEVIATION. An unintentional change of direction in a hole — "gone off".

DIAMOND BIT. A bit that uses diamonds as the cutting medium.

DIAMOND SIZE. The average number of diamonds to one carat in weight, usually expressed as a number of stones such as 8-12 or 40-60 stones per carat.

DIP METER. A portable device, usually battery operated, with an electrical probe on the end of a measured cable. Used for measuring water levels.

DIRECT CIRCULATION. Where water/mud is pumped down the drill-pipe and up the annulus between the drill-pipe and the side of the hole.

DIRECTIONAL DRILLING. The technique of controlled deflection of a well.

DISPLACEMENT CEMENTING. Where cement is forced into an annulus by pumping into casing, and then pumping mud onto a plug above the cement.

DOG-HOUSE. Office attached to major rig.

DOGLEG. A sharp change of direction in a well.

DOLLY. A device lowered into a well to facilitate the use of other equipment. For example, when measuring deviation.

DOPE. Grease used on drill-pipe/rod threads.

DOUBLE. A derrick that will rack drill-pipe/drill collars in pairs (stand) when tripping.

DOUBLE-ACTING. A reciprocating pump that pumps on both forward and reverse strokes.

DOWN TIME. Non-productive time, e.g. breakdown.

DOWN-THE-HOLE HAMMER. See hammer.

DRAG BIT. A bit with rigid blades or fingers.

DRAWDOWN. The difference between the static and dynamic water levels in a well.

DRAWWORKS. The main winch which lifts and lowers the drill string and off which drilling is done.

DRIFTER. A general name for an out-of-the hole hammer. A rotary percussive unit which exerts force from the rig by hammering onto drill-steels.

DRILLER. Operator of the rig.

DRILL-PIPE. The extension pipes added as drilling continues.

DRILL-RODS. The extension pipes added as drilling continues — core drilling.

DRILL-STEEL. Extension steels added as drilling continues — drifter drilling.

DRILL-TUBES. Lightweight drill-pipe mainly for down-the-hole hammer drilling.

DRILL COLLARS. Heavy walled drill-pipe-like tools used to add weight to the bit.

DRILL STEM. A heavy steel shaft behind the bit in cable tool drilling which exerts weight to the bit — similar to drill collars in concept.

DRILL STRING. The inhole tool assembly comprising drill-pipe, drill collars, bit, subs. etc.

DRILLING MACHINE. See rig.

DRILLING MUD. Usually a composition of one or more chemicals mixed with water which is pumped into the hole for cleaning, lubricating, hole stability etc. and circulated.

DRILLING SUPERINTENDENT. Will supervise a number of rigs.

DRIVE CLAMPS. Heavy duty clamps (see casing clamps) fitted to drill stem for driving casing.

DRIVE PIN. Used for connecting two joints (lengths) of augers which have union sockets.

DRIVE PIPE. Heavy duty casing.

DRUM. A container with a usable volume of 55 U.S. Gallons.

DUPLEX DRILLING. Simultaneous drilling and casing.

DUPLEX PUMP. A pump with two cylinders.

DYNAMIC LEVEL. The level at which the water settles during pumping.

ECCENTRIC BIT. A bit which is offset to one side to facilitate under-reaming.

EDUCTOR. A method of reverse circulation drilling where a jet of water effects the circulation.

ELEVATOR. A hinged circle or latched block used to lift coupled casing and externally upset drill-pipe.

EXTERNAL. UPSET. Where a tubular is thickened at its two ends for threading or welding. With external upset it is the outside diameter that is increased.

F.H. Full Hole — a thread form (API).

FACE DISCHARGE BIT. A coring bit which is drilled through the bit longitudinally so that flushing media discharges onto the cutting face.

FEED. Longitudinal movement of the drill string.

FEED PRESSURE. The amount of pressure in a hydraulic system registered in the feed circuit. Not to be confused with weight on bit.

FEED-OFF (TO). The action of lowering the tools whilst drilling.

FINGER BIT. See Drag Bit.

FINGER BOARD. That part of the system used for racking drill-pipe in the derrick.

FIR TREE BIT. A bit made up of ever increasing stages from a pilot to the final diameter of the hole.

FIRE-UP. To start, for instance, an engine.

FISH. An item of broken equipment or debris in a well.

FISHEYES. Globules of polymer created by trying to mix polymers too quickly. Polymers should, in powder form, be sprinkled onto briskly agitated water very slowly.

FISHING. The operation to recover lost tools or general debris from a well.

FISHING JAR. A long stroke jar used in the fishing string for driving in one direction.

FISHING MAGNET. A magnet which is lowered into a well for the recovery of small items of metallic trash left in the well. Can be solid state or electro-magnetic.

FISHING SPEAR. See fishing tap.

FISHING STEM. Relatively light, short stem used when carrying out fishing operations to provide alignment and some weight.

FISH TAIL BIT. A drag bit with two blades resembling the tail of a fish.

FISHING TAP. A tapered male fishing tool with hardened cutting threads cut into the outer face.

FISHING TOOL. A tool used for the recovery of debris (lost tools etc.) from the well.

FLAP VALVE. See Clack Valve.

FLASH WELDING. A method of securing tool joints to drill-pipe bodies by pressure and heat — no other metals added.

FLOCCULANT. A chemical to induce solids to attract in drilling muds, thereby leading to better "drop-out".

FLOCCULATION. The thickening of drilling fluids due to physical or chemical reaction.

FLOWMETER. Indicates the volume of fluids passed through its body, digitally.

FLUSH. To clean out a well. Also see Circulation.

FLUSH COUPLED. Casing that is joined by a coupling that is flush on the outside diameter with the casing. The inside diameter is slightly smaller.

FLUSH JOINTED. Casing that is joined together without a coupling and is flush on the outside diameter.

FLUTE. A groove in a cylindrical object, as in a fluted kelly.

FOAM DRILLING. Where hole-cleaning is done by foam powered by compressed air.

FOOT CLAMP. A clamp for drill-pipe/rods or small casing operated by the foot.

FOOT VALVE. A "clack" or "ball" or "seat" type valve situated at the bottom of an upstanding liquid pipe, such as the suction hose of a pump. Prevents loss of prime.

FRICTION WELDING. A method of securing tool joints to drill-pipe bodies by turning and heat — no other metals added.

FULL FACE BIT. A large diameter bit for drilling in a single pass.

GAUGE ANGLE. See Clearance Angle.

GAUGE RING. See Ring Gauge.

GAUGE WEAR. Wear on the extreme outside of the bit.

GOOSENECK. A bent pipe usually emanating from the mud swivel to which the mud hose is connected.

GRAVEL PACK. An array of sized, rounded gravel that is placed around well-screen to act as a filter, cleaning water from the aquifer as it enters the pump chamber.

GRINDING GAUGES. Gauges to check the result of bit grinding with special reference to clearance angle.

GROUTING. See Cementing.

GROUT MIXER. A machine to provide good quality mixing of grout materials.

GUAR GUM. A natural polymer used in low solids muds.

GUIDES. Fitted in the working table of a top drive rig and to the diameter of the tool (Collar etc.) in use, thus guiding that tool into the ground.

HAMMER. Generally, an air-operated percussive hammer. Can be mounted in the rig (Drifter) or will run in-hole (Down-the-Hole Hammer).

HAND. See Rig Hand. Also sometimes short for H-ave A N-ice D-ay.

HARD FACE. A hard, abrasion-resistant metal surface applied to softer metals.

HAWTHORN BIT. See Drag Bit.

HEC. Hydroxy Ethyl Cellulose.

HOISTING PLUG. A short threaded sub-attached to a bail via a thrust bearing used in tripping tubulars. Must have a relief hole drilled through.

HOLE-OPENER. A bit, with a pilot, or pilot bit in front, which increases the diameter of an existing hole.

HOLLOW STEM AUGERS. An auger built around a tube. The tube is drilled in with a recoverable bit in position then, with the bit removed, other works can take place.

IADC. International Association of Drilling Contractors.

ID. Inside diameter.

IF. Abbreviation of Internal Flush (thread form).

IGNEOUS ROCKS. Generally hard rocks, created from a molten mass that have cooled slowly, (examples are granite, diorite and gabbro), or quickly, as with basalt.

IMPREGNATED BIT. A bit faced with a mixture of powdered metal matrix and mesh-sized diamonds.

IMPRESSION BLOCK. A tube that is filled with wax (sometimes clay), fitted to the drill string, and pressed on bottom to take an impression of a fish.

INDUSTRIAL DIAMONDS. Those diamonds that cannot be used as gem-stones because of colour, shape or imperfections.

IN-HOLE. Signifies that an action is taking place down the hole.

INNER TUBE (BARREL). The inside tube of a double tube core barrel.

INSERT BIT. A bit which is inserted with hard metal such as tungsten carbide. This gives greater resistance to wear, therefore longer life.

INSIDE TAP. See fishing tap.

INTERNAL UPSET. Where a tubular is thickened at its two ends for threading or welding. With internal upset, it is the inside diameter that is reduced.

INTERNAL/EXTERNAL UPSET. Where a tubular is thickened at both ends for threading or welding. With internal/external upset, the inside and outside diameters are changed.

IRON. A euphemism for steel, as in "Big Iron", referring to drill collars which are heavy and made of steel.

JACK. A mechanical device for lifting heavy weights, can be hydraulically or mechanically (screw) actuated.

JACK HAMMER. Hand-held air-operated drill.

JARS. A link type device used in cable tool drilling to assist in freeing the bit in sticky conditions. There is also a rotary drilling jar for a similar duty.

JET MIXER. A device whereby mud (or cement) is mixed with water.

JET PIPE. Reverse circulation air lift drill pipe which allows air to pass into the column of water.

JETTING. A means of penetrating formations using high pressure fluids to remove cuttings.

JETTING TOOL. A device with a number of horizontal jets used for well development.

JOINT. A length of casing or drill-pipe etc.

JUNK. Foreign matter inhole.

JUNK BASKET. A perforated tube run above the bit to retain junk.

JUNK MILL. A special bit with many rows of teeth for drilling out junk.

KALGON. A proprietary name for a polyphosphate used for cleaning out mud residue from a well during development. Sometimes spelt CALGON.

KAOLINITE. A clay used in mud drilling.

KELLY (BAR). A square, hexagonal, fluted (etc.) tubular driven by the rotary table, thus transmitting rotary motion to the drill string.

KELLY BUSHINGS. Bushings in the rotary table that match the configuration of the Kelly, thus transmitting rotation.

KELLY DRIVE. The mechanism for driving the Kelly.

KERF. The width of the drilling face of a coring bit, not the bit diameter.

KICK. A sudden rising motion in the mud pits. This is a sign of excessive in-hole pressure and indicates danger.

KILL. To overcome excessive formation pressures in a wild well.

KING BOLT. A large bolt holding the upper end of sheer legs together and from which the sheave is suspended.

L/Sec. Litres per second of compressed air.

LANDING. A point in a well where casing is to be set.

LANDING JOINT. A special length of casing, usually equipped with cementing head, and other affiliated tools and used to accurately place casing for cementing inhole.

LATCH JACK. A fishing tool for retrieving an item with a bail atop.

LAYKEY. A spanner type device for holding drill-pipe which has "flats" machined in the tool joints, in the table.

LAZY SUSAN. See Carousel.

LCM. See Lost Circulation Material.

LEFT HAND THREADS. A thread which tightens in an anti-clockwise rotation.

LINER. The sleeves in a pump (mud-pump for instance) inside which the pistons travel.

LOST CIRCULATION MATERIAL (LCM). Materials used in a circulating system to plug porous formations, thus preventing Lost Circulation or Mud Losses.

LOST CIRCULATION. When the circulating fluid fails to return to surface due to broken or cavernous formations.

LOW SOLIDS MUDS. The use of natural and/or synthetic polymers instead of clay.

LUBE. A lubricant introduced into the well to reduce friction between the drill string and the walls of the well.

MACHINE THAT DRILLS. An imposter that makes some holes, but inefficiently.

MAGNET. See Fishing Magnet.

MAKING-UP. Putting on the "next" joint of drill-pipe etc.

MALE THREAD. Pin.

MANOMETER. A gauge.

MARSH FUNNEL AND CUP. A funnel shaped device with a measured volume and a specific exit orifice through which the viscosity of mud is measured relative to water.

MAST. See Derrick.

MATRIX. The metal bond in which diamonds and tungsten carbide are held. Also related to boulder/gravel formations which are usually held together by sand etc.

MESH. A sieve analysis. Chip samples can be put through a series of vertically stacked sieves, each progressively smaller, thus each will retain a size (mesh).

METAMORPHIC ROCKS. See Metamorphosis.

METAMORPHOSIS. A process by which a rock has been changed through heat and pressure. For instance, shales become slate.

MILLING CUTTER. See Junk Mill.

MONKEY BOARD. A raised platform in the derrick used when running tools such as drill-pipe and casing.

MONTMORILLONITE. A clay used in mud drilling.

MOUSEHOLE. On a major rig, a pre-drilled hole in which the "next" joint of drill-pipe is parked.

MUD AGITATOR. Keeps mud in good condition by a stirring motion.

MUD BALANCE. Used for checking mud weights and specific gravity.

MUD DRILLING. When hole cleaning etc., is actuated by a drilling mud. See Circulation.

MUD ENGINEER. A very experienced engineer who will control the mud on a major rig.

MUD GUN. A jet-like device fitted to the washdown line from the mud pump and used for agitating the mud and "washing down".

MUD LOSSES. When porous formations absorb a part of the circulation fluid.

MUD MIXER. See Jet Mixer.

MUD PITS. When mud drilling, two or more pits are used for cleaning the mud and these have enough volume to ensure the mud is clean during circulation.

MUD SCOW. A heavy duty bailer fitted with cutting shoe as well as a "clack-valve" or similar as in cable tool drilling.

MUD TANK. Where mud drilling tanks are used as the reservoir. The rig must have sufficient height under the table to facilitate the use of tanks.

MUD UP. Either to go from water to mud circulation or to increase the density of the mud.

NOMINAL SIZE. When dimensions are quoted to the nearest one sixteenth of an inch.

NOTCH. See Weir, but where a weir is usually rectangular, a notch is in the shape of a vee, hence vee-notch.

OD. Outside Diameter.

OFF BOTTOM. When the bit is held away from the bottom of the well.

OIL BASED MUD. Where oil is used as the basis of the mud rather than water. Can also be applied when high percentages of diesel oil are added to mud.

ON BOTTOM. When the bit is in contact with the drilling face.

OPEN HOLE. Normal drilling as opposed to core drilling.

ORGANIC FORMATIONS. Remains of plant life that have undergone chemical and pressure changes. Examples would be peat, lignite and coal.

OUTER TUBE. The outer tube of a multi-tubed core barrel.

OVERPULL. The amount of "pull" in excess of a given amount — i.e. if, with ten tons of tools in-hole you are pulling twelve tons, then you have two tons overpull.

OVERSHOT. In fishing this is a female fishing tool. In wire-line coring, it is the tool that runs down the drill rod to recover the inner barrel.

PAC. Poly Anionic Cellulose.

PPL. A plastic (polypropelene) used in some casing manufacture.

PVC. A plastic (poly vinyl chloride) used in some casing manufacture.

PACKER. A device by which a selected zone in a well can be isolated from others, temporarily, for testing that zone for, say, permeability.

PARMALEE WRENCH. A wrap around wrench especially suitable for dismantling core barrels but excellent for others uses on tubulars.

PASS. A drilling pass; e.g. a single pass is drilling the hole in one go.

PERIPHERAL SPEED. Related to rotational speeds, this is the time taken for a point on the circumference of a bit to travel back to the same point, in feet per minute.

PERMEABILITY. The measurement of water flow into given formations at various pressures.

PIEZOMETER. An instrument installed in a well that measures variations in ground-water.

PIG. A drillable plug placed between cement and mud when cementing by the displacement method.

PILOT. A guide that is installed ahead of a bit to act as a guide to the bit. Runs in an existing hole known as a pilot hole or, sometimes, a rathole.

PILOT BIT. See Pilot, but in this case is a bit. Sometimes refers to the lowest bit in a hole-opener or Fir Tree Bit.

PILOT HOLE. A slim hole drilled ahead of the final diameter to — say — investigate hazards or to check the water table, etc.

PIN. The male thread.

PIPE WIPER (ALSO ROD WIPER). A tight flexible gland through which drill-pipe or rods pass when being tripped, thus being cleaned of mud etc.

PIPE. A shortening of either drill-pipe or casing-pipe.

PLUG-IT. A proprietary lost circulation material.

POLY ANIONIC CELLULOSE. A synthetic polymer used in low solids muds.

POLYPHOSPHATES. A collective name that includes those chemicals used in well development for cleaning out mud residue.

POWER TONG. A mechanically actuated tong for making up or breaking out tubulars.

PRESSURE RELIEF VALVE. A pre-set safety valve which will "blow off" at the pre-set pressure thus preventing undue stresses on, say, a mud pump.

PRIME MOVERS. The engines/motors etc. that power a rig.

PULLDOWN. Exerting pressure on the drill string from the top. The use of drill collars is far superior.

PULSATION DAMPER. A bottle-like device that absorbs the pulsations from a reciprocating type pump, thus giving a smooth delivery into the standpipe.

PUMPING TEST. A controlled test to check the volume of water available from a water well.

PUP JOINT. A short length of tubular, e.g. casing or drill-pipe.

RANDOMS. Sometimes casing and drill-pipe are produced in varying lengths in a single batch; these are random lengths.

RATHOLE. On a major rig this is a pre-drilled hole in which the Kelly is "parked" during non-drilling operations, when a trip is in progress, or when making a connection. It can also refer to a pilot hole.

REAM. To enlarge the diameter of a well.

REAMER. A device run behind the bit — usually hard faced or tungsten carbide inset — that maintains the diameter of a well, thus compensating for bit gauge wear.

REAMING BIT. A bit used to enlarge the diameter of a well.

RECOVERY. The time taken for water to travel from the dynamic level to the static water levels in a well when pumping has stopped.

RECOVERY. The amount of core recovered relative to the amount drilled (%).

REG. Abbreviation of Regular (thread form).

RELIEF VALVE. See Pressure Relief Valve.

REPLACEABLE BLADES. A type of drag bit where the blades are replaceable when worn or when different diameters are necessary.

RETURNS. Cuttings coming to surface in the flushing medium.

REVERSE CIRCULATION. Where water/mud passes down the annulus between the drill-pipe and the side of the hole and is drawn up inside the drill-pipe.

RHEOLOGY. The study of fluids in motion.

RIG. The drilling unit as a whole, less tools.

RIG DOWN. To prepare a rig for moving at the end of a well.

RIG HAND. Somebody who works on a rig.

RIG UP. To prepare a rig for drilling.

RING GAUGE. A steel plate with a hole cut in it conforming to a bit diameter, used to check wear and that one bit will follow another in a hole.

ROCKBIT. A bit with rolling cones.

ROCKWELL. A measurement of hardness expressed in the Rockwell scale.

ROTARY HEAD. Unit that rotates the drill string from the top.

ROTARY TABLE. Unit that rotates the drill string via a Kelly-bar.

ROTATION SPEED. The speed at which the bit is turned. Measured in RPM (revolutions per minute).

ROUGHNECK (FLOOR MAN). A drill helper who will work around the table.

ROUND TRIP. To recover from the hole and then return the drill string. To — say — change the bit.

ROUSTABOUT. A labourer around the rig.

RPM. Revolutions per minute.

RUN. The action of placing tools in a well.

SPM. Strokes per minute of a reciprocating pump.

SADDLE. A "U" shaped device into which a bit roller is pinned, mainly used in building large diameter full face bits or hole openers.

SAFETY HOOK. A lifting hook that has a spring to prevent the load leaving the hook.

SAFETY LATCH. An elevator or other such device where a latch is installed to prevent premature unlatching.

SAMPLE. A sample of the cuttings from the formations being drilled.

SAND LINE. A thin wire rope used to raise and lower the bailer or sand pump when cable tool drilling.

SAND PUMP. A piston type bailer.

SAND TRAP. A device for separating coarser materials from the fines.

SAWTOOTH BIT. A core bit or casing shoe which has a serrated cutting face.

SCREEN. A tubular which is perforated in varying degrees to allow the passage of water from the aquifer into the pump chamber, but blocking access of solids.

SCREEN SIZE. The measurement of the gap in a screen — see SCREEN. Also see Mesh.

SCREW PUMP. A type of mud pump which is a screw inside a stator.

SEDIMENTS. Formations that have been "laid down" over millions of years. Examples would be limestone, sandstone and shale.

SERVICE WINCH. Usually associated with top drive rigs — a winch for casing handling and other duties.

SET BIT. A core bit that is set with whole diamonds, not diamond powder.

SETTING CHARGE. The charge made for setting a bit with diamonds, but excluding the cost of the diamonds.

SETTLING PIT. See Settling Tank.

SETTLING TANK. The first of two (or more) tanks in a circulating system encountered by the mud. All remaining cuttings in the mud should have been dropped here.

SHALE SHAKER. A mechanically vibrated sieve which cleans mud from the well.

SHALE TRAP. A diaphragm type of device put around casing at a pre-determined level onto which cementing is done.

SHEAVE. A grooved wheel which holds wire-line or rope around its circumference.

SHOCK ABSORBER. Associated with down-the-hole hammers. It is fitted on top of the hammer to absorb shock away from the drill string.

SIDE-ARM LOADER. Lifts drill-pipe from ground level into the mast for make-up. Top drive rigs only.

SIDETRACK. To bypass an obstruction in a well.

SINGLE. A derrick that will rack only one joint of drill-pipe when tripping.

SINGLE-ACTING. A reciprocating pump that pumps on the forward stroke only.

SINKER BAR. A heavy duty rod used in cable tool drilling.

SKID. A sled-like mounting for small rigs, pumps etc.

SLIDING JARS. See Jars.

SLIM HOLE. A term to differentiate small diameter drilling for oil with normal oil field diameters; it is drilled at plus or minus 5 inches.

SLIP ROPE. Used in conjunction with a light winch such as a cathead. One end is held by the operator.

SLIP-PLATE. See Bitch.

SLIPS. Tapered serrated wedges working inside a similarly tapered bowl (or spider) which hold drill-pipe and casing in the table.

SLIPS BOWL. See Slips.

SLOT SIZE. The gap in well-screen through which water passes.

SLOUGH. Swelling and collapsing associated with shales when wet.

SNATCH BLOCK. A single sheaved travelling block with side opening.

SODIUM HEXAMETAPHOSPHATE. A chemical used for cleaning out mud residue in a well during development.

SPIDER. See Slips.

SPOOL. A pair of flanges separated by a steel tube from which emanates a number of valves. Used in blow-out-preventer stacks.

SPUD (TO). See Spudding.

SPUDDING. To start a well — Spud In.

SPURT LOSS. When there is a partial separation of water from the chemicals in a circulating system and that water is absorbed into the formations.

STABILISER. A device fitted in several places in a string of drill collars, taking up almost the diameter of the well, thus giving support and rigidity, but passing fluids.

STACK. A term relating to blow-out-preventer equipment on a well comprising bop(s), spool, flanges etc.

STANDPIPE. The rigid pipe running up the derrick with a flexible hose from the mud pump at the bottom and another connecting to the swivel at the top.

STANDS. Drill-pipe or collars stacked vertically in a derrick. Can be one, two, three or even four (fourble) joints per stand.

STATIC LEVEL. The rest water level in a well — the pump would be switched off.

STONE TRAP. A vacuum chamber situated between the drill-pipe and the suction pump in reverse circulation to drop out large particles, thus saving the pump.

STRAINER. Filters out debris from a suction hose where it is fitted at the entry from the pits (tanks).

STRING. See Drill String.

STROKE. The distance travelled by a piston in a liner.

STUCK. The tools are trapped in the well.

STUFFING BOX. A chamber with packings which forms a seal around, for instance, piston rods in a mud pump, drill-pipe/rods at the top of the hole.

SUB. See substitute.

SUBSTITUTE. An adaptor from one thread to another, or a wearing item.

SUBSTRUCTURE. A rig that is built high off the ground on a structure to give sufficient head room for blow-out-preventers, etc.

SUCTION HOSE. The hose through which mud is drawn from the pits.

SUCTION PIT. See Suction Tank.

SUCTION TANK. The last tank in the circulating system from which clean mud is pumped into the well.

SURGE BLOCKS. A device used in well development for settling the pack around the screen.

SWAB (TO). When the drill string is pulled too rapidly, leading to hole collapse caused by a piston-effect on unstable formations.

SWAGE. A conical adaptor.

SWIVEL. A device comprising two cylinders separated by packings through which drilling fluids pass. The out cylinder remains still and the inner rotates.

TPI. Abbreviation of Threads Per Inch.

TAPERED THREAD. A conical thread.

TD. Total depth.

TETRASODIUM PYROPHOSPHATE. A chemical used for cleaning out mud residue in a well during development.

TFA. Total flow area. The area of a diamond full face bit allowed for the passage of drilling fluids.

THIN-WALLED BIT. A core bit with a narrow kerf.

THIXOTROPY. The peculiarity of some substances to be thick when motionless and fluid when agitated. An example would be Tomato Ketchup.

THREAD PROTECTOR. A steel or plastic cap or plug for the protection of threads when not in use.

THRIBBLE (TRIPLE). A derrick that will rack drill-pipe (and collars) in stands of three when tripping.

TOLERANCE. The dimensional limits in manufactured items and the ability to be reasonable. The latter is known to be exponentially increased by the consumption of tea in large volumes.

TONGS. The collective name for tubular handling wrenches.

TOOL. A collective name meaning anything that does something, e.g. bits, subs., spanners, etc. etc.

TOOL PUSHER. A foreman driller who will supervise the rig throughout the day and night shifts.

TOOL-JOINTS. Those threaded units at each end of drill-pipe.

TORQUE. The effort required to turn an object against a resistance.

TORQUE GAUGE. Indicates the amount of torque being absorbed by the bit.

TOTAL DEPTH. See TD.

TPI. Threads per inch.

TRAMP IRON. See Junk.

TRANSFER CASE. A type of gearbox that transfers power from an input shaft to two or more output shafts.

TRASH. Another word for Junk.

TRAVEL. The linear movement of a component, e.g. the stroke of a piston.

TRAVELLING BLOCK. A moving sheave block e.g. the block that carries the drill string on its hook in a rotary table rig.

TREMIE PIPE. A pipe through which cementing is done.

TRI-CONE BIT. A registered trade name owned by The Hughes Tool Co.

TRIP. To recover the drill string from the well.

TRIPLEX PUMP. A pump with three cylinders.

TRIPOD. Three legged derrick.

TUBULARS. Anything that is tubular, such as casing, tubing, pipe etc.

TUNGSTEN CARBIDE INSERT. Wear resistant inserts of tungsten carbide inserted into some forms of bit. Also applied to other wearing items such as reamers.

TUNGSTEN CARBIDE. An extremely hard metal.

TURBO DRILLING. Where drilling mud drives a down-the-hole motor which in turn drives the bit; the drill-pipe stays still.

TWIST-OFF. The shearing of drill-pipe/rods/collars inhole.

UNDER-REAM. To increase the diameter of a well below the casing.

UNDER-REAMER. A bit for under-reaming; see Under-ream.

UNIVERSAL JOINT. Articulated joints used to transmit shaft drive and permitting some misalignment.

UPHOLE. A seismic term for a hole which is drilled normally but blasted from the bottom up.

UPHOLE VELOCITY. The speed at which the flushing medium flows upwards in the well.

UPSET. See Internal Upset, External Upset and Internal/External upset.

VEE NOTCH, (Or weir). Can be either 60 degrees or 90 degrees, and is used for measuring water flow during pump testing.

WASH-OVER STRING. A string of tools which can be lowered over a stuck drill string to drill out the obstruction.

WASHBORING. Drilling with a high pressure jet of water.

WASHOUT. Where a soft layer is eroded by the flushing medium.

WATER WELL. A hole drilled in the ground for the abstraction of water.

WEAR SUB. A threaded piece that is put between large wearing parts to take the wear away from those parts.

WEDGE. A long tapered tool inserted into a well against which to deflect the direction of the well.

WEIGHT INDICATOR. A gauge that indicates the weight on bit and, often, the drill string weight as well.

WEIGHT ON BIT. The amount of drill collar weight being applied to the bit.

WEIR. A device over which water flow is measured during development of a well.

WEIR TANKS. Used for settling and measuring water during pump testing.

WELDING JIG. A device like a casing clamp that has openings through which casing can be tack welded prior to the main weld — it helps to achieve straightness.

WELL-SCREEN. See Screen.

WIPER TRIP. To run the tools into a hole to check for settlement, collapse and general integrity of the well.

WIRE-LINE CORING. Where the outer core barrel remains in the hole and the inner barrel is recovered for removal of the sample by an overshot. See Overshot.

WIRE-LINE. Wire ropes.

WORKOVER. The process of working inside an existing well, perhaps to change the pump, or to clean out etc. etc.

X BIT. A percussive bit with four cutting edges in the shape of an "X".

XANTHAN GUM. A polymer mud additive.

Index

The first number gives the book, followed by the page number and illustration letter.

For Product Safety Concerns and Information please contact our EU
representative GPSR@taylorandfrancis.com
Taylor & Francis Verlag GmbH, Kaufingerstraße 24, 80331 München, Germany

www.ingramcontent.com/pod-product-compliance
Ingram Content Group UK Ltd.
Pitfield, Milton Keynes, MK11 3LW, UK
UKHW051836180425
457613UK00023B/1273